MATHS PLUS

MENTALS AND HOMEWORK BOOK

AUSTRALIAN CURRICULUM

Harry O'Brien
Greg Purcell

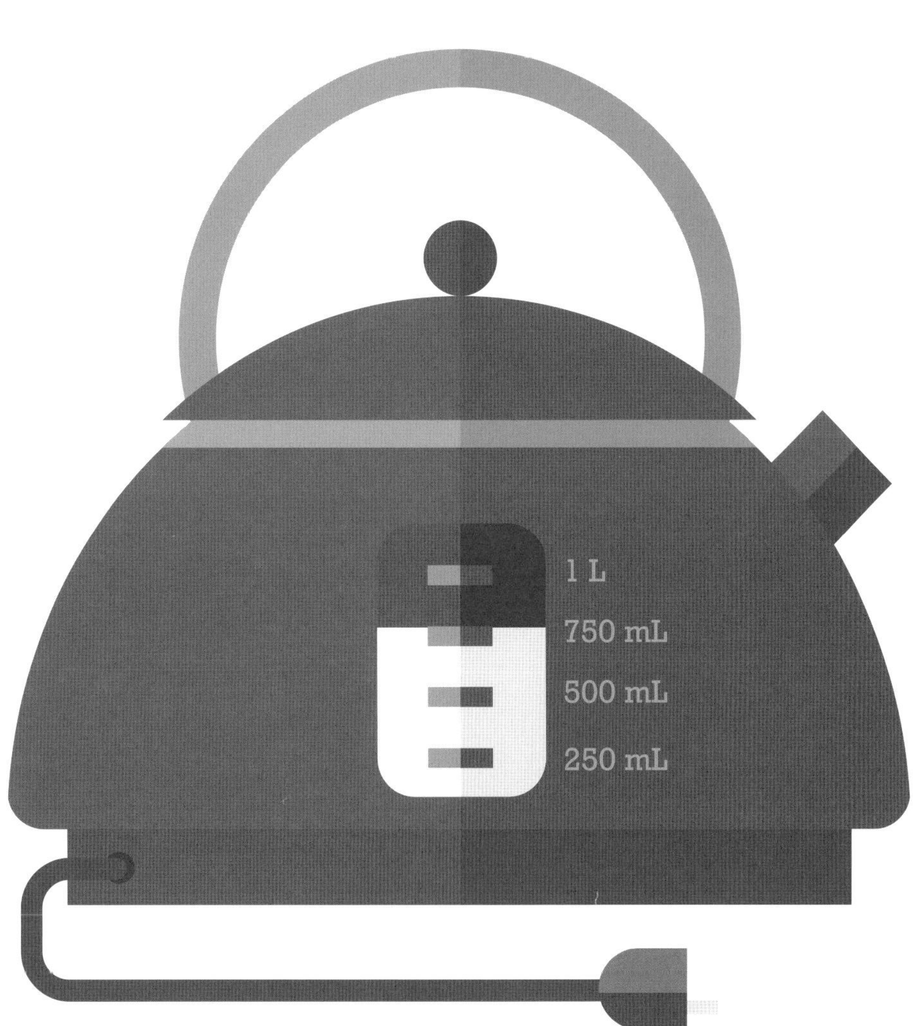

Contents

Contents

Australian Curriculum AC Linking Chart

Units	1	2	3	4	5	6	7	8	9	10
Number and Algebra										
Recognise, represent and order natural numbers using naming and writing conventions for numerals beyond 10 000 (AC9M3N01)					✓	✓				
Recognise and represent unit fractions including $\frac{1}{2}, \frac{1}{3}, \frac{1}{4}, \frac{1}{5}$ and $\frac{1}{10}$ and their multiples in different ways; combine fractions with the same denominator to complete the whole (AC9M3N02)								✓		
Add and subtract two- and three-digit numbers using place value to partition, rearrange and regroup numbers to assist in calculations without a calculator (AC9M3N03)			✓	✓	✓		✓	✓	✓	✓
Multiply and divide one- and two-digit numbers, representing problems using number sentences, diagrams and arrays, and using a variety of calculation strategies (AC9M3N04)		✓	✓	✓		✓	✓			
Estimate the quantity of objects in collections and make estimates when solving problems to determine the reasonableness of calculations (AC9M3N05)										
Use mathematical modelling to solve practical problems involving additive and multiplicative situations including financial contexts; formulate problems using number sentences and choose calculation strategies, using digital tools where appropriate; interpret and communicate solutions in terms of the situation (AC9M3N06)				✓	✓					
Follow and create algorithms involving a sequence of steps and decisions to investigate numbers; describe any emerging patterns (AC9M3N07)		✓			✓					✓
Recognise and explain the connection between addition and subtraction as inverse operations, apply to partition numbers and find unknown values in number sentences (AC9M3A01)	✓		✓	✓						✓
Extend and apply knowledge of addition and subtraction facts to 20 to develop efficient mental strategies for computation with larger numbers without a calculator (AC9M3A02)	✓	✓	✓	✓	✓			✓	✓	✓
Recall and demonstrate proficiency with multiplication facts for 3, 4, 5 and 10; extend and apply facts to develop the related division facts (AC9M3A03)		✓	✓	✓		✓	✓			
Measurement and Space										
Identify which metric units are used to measure everyday items; use measurements of familiar items and known units to make estimates (AC9M3M01)	✓			✓			✓			✓
Measure and compare objects using familiar metric units of length, mass and capacity, and instruments with labelled markings (AC9M3M02)	✓						✓			✓
Recognise and use the relationship between formal units of time including days, hours, minutes and seconds to estimate and compare the duration of events (AC9M3M03)									✓	
Describe the relationship between the hours and minutes on analog and digital clocks, and read the time to the nearest minute (AC9M3M04)									✓	
Identify angles as measures of turn and compare angles with right angles in everyday situations (AC9M3M05)										✓
Recognise the relationships between dollars and cents and represent money values in different ways (AC9M3M06)										
Make, compare and classify objects, identifying key features and explaining why these features make them suited to their uses (AC9M3SP01)	✓			✓				✓		
Interpret and create two-dimensional representations of familiar environments, locating key landmarks and objects relative to each other (AC9M3SP02)			✓		✓					
Statistics and Probability										
Acquire data for categorical and discrete numerical variables to address a question of interest or purpose by observing, collecting and accessing data sets; record the data using appropriate methods including frequency tables and spreadsheets (AC9M3ST01)		✓	✓			✓			✓	
Create and compare different graphical representations of data sets including using software where appropriate; interpret the data in terms of the context (AC9M3ST02)			✓			✓			✓	
Conduct guided statistical investigations involving the collection, representation and interpretation of data for categorical and discrete numerical variables with respect to questions of interest (AC9M3ST03)			✓							
Identify practical activities and everyday events involving chance; describe possible outcomes and events as 'likely' or 'unlikely' and identify some events as 'certain' or 'impossible' explaining reasoning (AC9M3P01)										
Conduct repeated chance experiments; identify and describe possible outcomes, record the results, recognise and discuss the variation (AC9M3P02)										

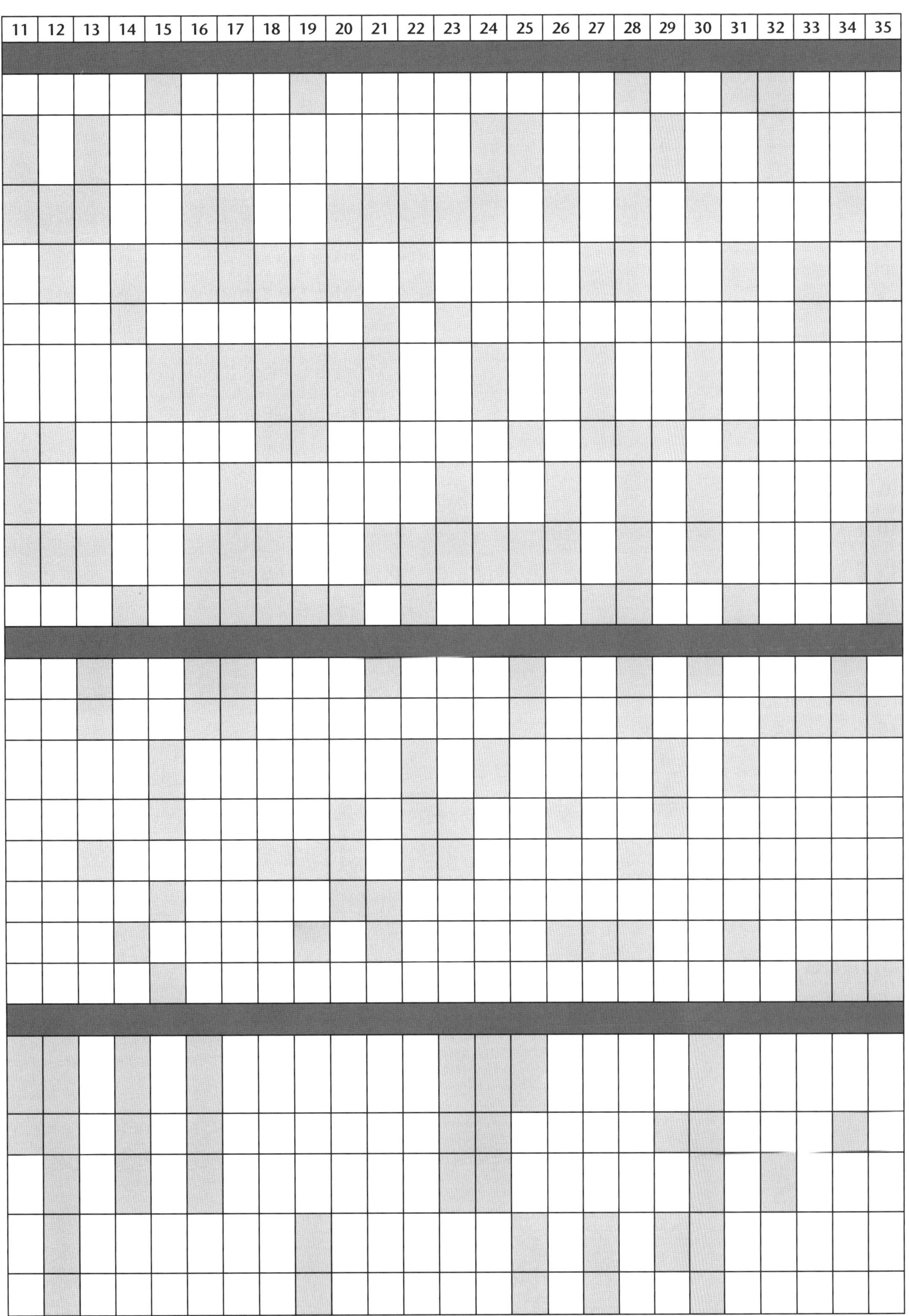
11 12 13 14 15 16 17 18 19 20 21 22 23 24 25 26 27 28 29 30 31 32 33 34 35

UNIT 1

Number and Algebra

SET 1 Basic

1 6 + 2

2 6 + 4

3 2 + 2

4 5 + 4

5 7 + 1

6 8 + 3

7 9 + 10

8 2 + 6

9 8 − 3

10 7 − 1

11 3 − 3

12 10 − 2

13 2, 4, 6, ☐

14 1, 3, 5, ☐

15 I had a dozen eggs but I dropped 7. How many eggs were left?

☐ eggs

SET 2 Addition facts

1 Add 5 and 4.

2 Sum of 9 and 8

3 6 plus 5

4 What is the sum of 9 and 7?

5 What is 5 more than 13?

6 What is 6 more than 7?

7 What is 9 plus 7?

8 13 + 0

9 9 + ☐ = 15

10 Add $7 to $13.

11 11 cm + 5 cm = ☐ cm

12 11 plus 8

13 What number added to itself makes 18?

14 John spent $9 and $5. How much did he spend in total?

15

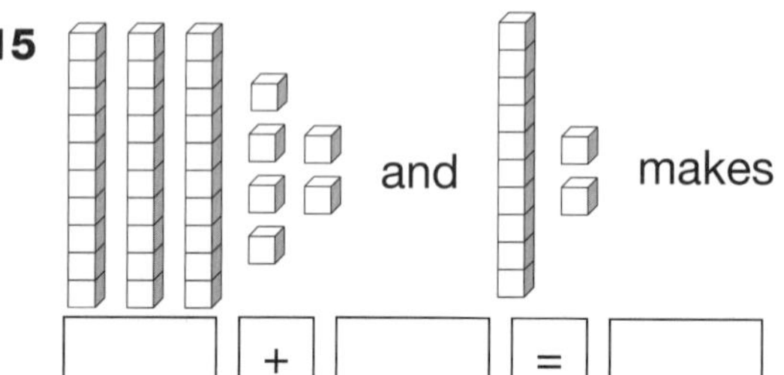

☐ + ☐ = ☐

Space Three-dimensional objects

Match the named 3D shapes to their environmental shapes.

1

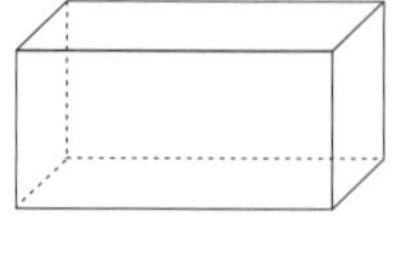

prism

2

cone

3

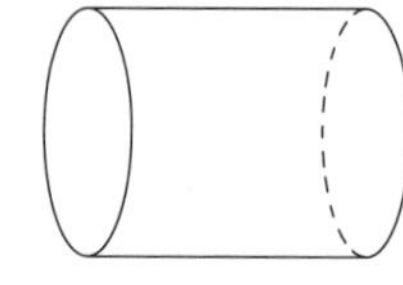

cylinder

Number and Algebra

SET 3 Skip counting

Complete the skip counting patterns.

1	0	3	6	9			
2	12	14	16	18			
3	40	50	60	70			
4	0	5	10	15			
5	6	10	14	18			
6	20	23	26	29			
7	30	35	40	45			

Complete each pattern and then write the rule.

8	50	55	60	65			

Rule: ______________________

9	36	34	32	30			

Rule: ______________________

10	21	26	31	36			

Rule: ______________________

SET 4 Extension

1 55 = ☐ tens + ☐ ones.

2 Which is smaller: 17 + 6 or 28 + 4?

3 $9 + $15

4 5 tens + 6 ones

5 Write 15 in words. ☐

6 $3 + $13 + $22

Shade two numbers in each circle that total 27.

7

15, 20, 16, 3, 13, 7, 14, 12

8

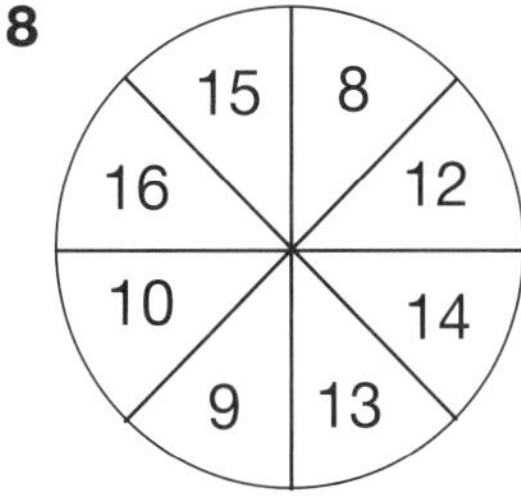

Mathematical Reasoning

Find 4 pairs of numbers that have a total of 20.

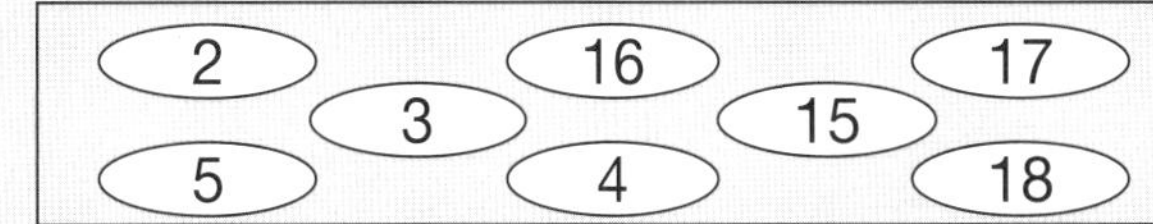

9 __________ + __________ = 20

10 __________ + __________ = 20

11 __________ + __________ = 20

12 __________ + __________ = 20

Measurement Centimetres

Use the 1-centimetre dot paper to draw the following lines. The first one has been done for you.

5 cm

6 cm

12 cm

15 cm

13 cm

UNIT 2

Number and Algebra

SET 1 Basic

1 7 + 2

2 6 + 4

3 5 + 6

4 9 + 2

5 9 + 4

6 5 + 6

7 3 + 7

8 10 – 5

9 7 – 4

10 9 – 6

11 12 – 3

12 Half of 10

13 Double 6.

14 7 + ☐ = 10

15

SET 2 Subtraction facts

1 Take 10 from 20.

2 Subtract 5 from 17.

3 Subtract 8 from 15.

4 Take $4 from $13.

5 What is the difference between $7 and $19?

6 What is the difference between 7 and 4?

7 13 km less than 20 km.

8 Take 14 cm from 64 cm.

9 Subtract 6 ones from 2 tens.

10 15 less than 19

11 Take 11 minutes from 20 minutes.

12 Subtract 7 from 13.

13 By how much is 13 less than 18?

14 Deduct 1 from 12.

15 Deduct 13 pages from 17 pages.

16 Subtract 4 from 20.

Number and Algebra Repeated addition

How many counters are in each set?

1 ○○ ○○
○ ○

2 groups of 3

3 + 3 = ☐

2 ○○ ○○ ○○ ○○
○ ○ ○ ○

4 groups of 3

☐ + ☐ + ☐ + ☐ = ☐

3 ○○ ○○ ○○ ○○
○○ ○○ ○○ ○○

4 groups of 4

4 + ☐ + ☐ + ☐ = ☐

4 ○○ ○○ ○○ ○○ ○○
○○ ○○ ○○ ○○ ○○

5 groups of 4

☐ + ☐ + ☐ + ☐ + ☐ = ☐

Number and Algebra

SET 3 Odd and even numbers

Model these numbers on the dot paper. Then tick whether they are odd or even. The first one is done for you.

	Number	Odd	Even
1	7	✔	
2	10		
3	12		
4	17		

Write odd or even after the following numbers.

5 23 ____________

6 44 ____________

7 63 ____________

8 88 ____________

9 96 ____________

10 123 ____________

11 134 ____________

12 139 ____________

13 486 ____________

SET 4 Extension

1 16 – 12

2 48 – 6

3 15 minus 6

4 Eighteen take away 6

5 What season comes after spring?

6 How many days in May?

7 When is Christmas Day?

8 If Monday was 3 June, what would be the date of the next Wednesday?

9 How many 10c coins in $2?

10 What number is between 66 and 68?

11 Share 12 lollies between 4.

12 15 + 6 – 2

13 14 km minus 4 km

14 13 + 8 – 1

15 Write 3 number sentences that have an answer of 12.

☐ – ☐
☐ – ☐ = 12
☐ – ☐

Statistics Column graphs

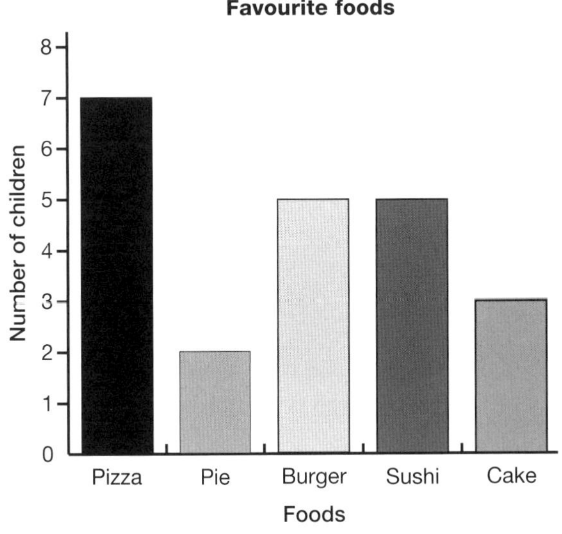

1 How many children liked sushi the best? ☐

2 What is the most popular food? ☐

3 What is the least popular food? ☐

4 How many children like cake the best? ☐

5 How many children like pizza the best? ☐

6 Which two foods were liked by the same number of children? ☐

UNIT 3

Number and Algebra

SET 1 Basic

1 6 + 5

2 4 + 3

3 9 + 2

4 5 + 4

5 6 + 0

6 8 + 5

7 9 − 5

8 8 − 7

9 10 − 10

10 10 − 4

11 7 take away 3

12 10 take away 9

13 Half of 12

14 Double 7.

15 Jim ate 15 peanuts and Lauren ate 7. How many peanuts did they eat altogether?

☐ peanuts

SET 2 Equal groups/ multiplication

Write number sentences to describe the groups.

1 [2] groups of ☐ = ☐

2 ☐ groups of ☐ = ☐

3 ☐ groups of ☐ = ☐

4 ☐ groups of ☐ = ☐

5 ☐ groups of ☐ = ☐

Space Describing position

Petri built a model out of blocks. Put the letters on the blocks to discover the secret message.

1 Put a **C** on the top left block.

2 Put an **N** on the middle block.

3 Put a **D** on the bottom left block.

4 Put a **T** on the top right block.

5 Put a **G** on the bottom right block.

6 Put another **D** to the right of **N**.

7 Put an **A** to the left of **N**.

8 Put an **O** between **D** and **G**.

9 Put another **A** between the **C** and **T**.

10 What is the message?

____

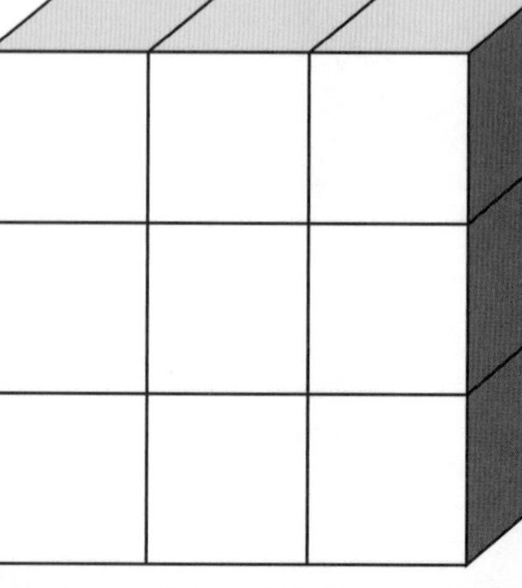

Number and Algebra

SET 3 Connecting addition/subtraction

Check these addition facts using subtraction.
Put a tick (✓) for correct facts and a cross (✗) for incorrect ones.

				✓ or ✗
1	5 + 8 = 13	➔ 13 – 8 =	☐	☐
2	12 + 9 = 20	➔ 20 – 9 =	☐	☐
3	7 + 11 = 18	➔ 18 – 11 =	☐	☐
4	8 + 6 = 14	➔ 14 – 6 =	☐	☐
5	13 + 6 = 18	➔ 18 – 6 =	☐	☐

Check these subtraction problems using additions.

6 A golfer had 12 balls and lost 7 in the scrub. How many balls are left?

☐ – ☐ = ☐ ☐ + ☐ = ☐

7 Jacqui picked 18 strawberries. If 5 were rotten, how many could she eat?

☐ – ☐ = ☐ ☐ + ☐ = ☐

8 Tom baked 38 loaves of bread but sold only 25 loaves. How many were left? Solve the question and write an inverse operation to check the answer.

☐ – ☐ = ☐ ☐ + ☐ = ☐

SET 4 Extension

Write number sentences to describe the groups.

1 69 = ☐ tens + ☐ ones.
2 Which is greater: 17 + 9 or 8 + 10?
3 Forty-five and six more
4 49 + 18
5 How many eyes would 8 children have?
6 19 = ☐ ten and ☐ ones.
7 Which is the larger, 8 tens and 9 ones or 9 tens?
8 25 plus 29
9 What number is between 40 and 42?
10 Which season comes after summer?
11 38 + 9
12 36 = ☐ tens + ☐ ones
13 50, 55, ☐, 65, ☐
14 3 tens + 8 ones

Mathematical Reasoning

Shade two numbers in each circle that multiply to give 24.

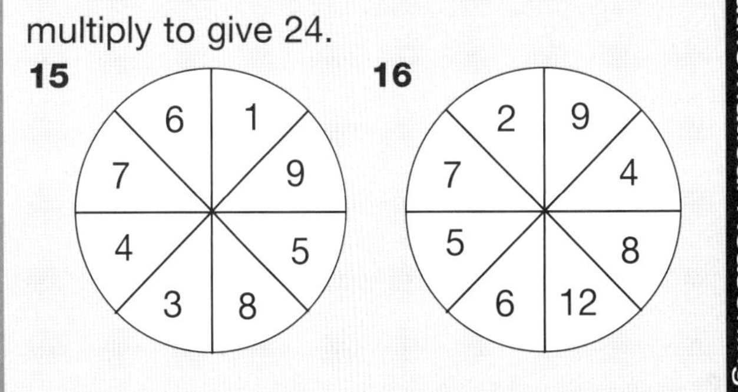

Statistics Collecting and displaying data

Teachers' cars

Toyota	|||
Kia	𝍸 𝍸 |||
Ford	𝍸 𝍸
Mazda	|||
Volkswagen	||
Mitsubishi	𝍸 |

1 How many teachers drive Fords?
2 How many drive Kias?
3 How many drive Toyotas?
4 Which is the most popular car at this school?

☐

5 Which cars have equal popularity?

☐

UNIT 4

Number and Algebra

SET 1 Basic

1 10 + 4

2 7 + 6

3 7 + 7

4 8 + 5

5 2 + 12

6 13 + 4

7 8 + 7

8 9 + 5

9 11 – 2

10 13 – 2

11 14 – 2

12 16 – ☐ = 14

13 Double 8.

14 Half of 16

15

Susan had $6 but Peter had twice as much. How much money did Peter have?

$ ☐

SET 2 Inverse operations

Complete these addition facts. Then use the subtraction facts to check your answer.

1	6 + 7 = ☐	13 – 6 = ☐
2	8 + 6 = ☐	14 – 8 = ☐
3	10 + 9 = ☐	19 – 10 = ☐
4	12 + 5 = ☐	17 – 12 = ☐
5	3 + 13 = ☐	16 – 3 = ☐
6	18 + 2 = ☐	20 – 18 = ☐
7	9 + 9 = ☐	18 – 9 = ☐

8 Lisa cooked 18 biscuits. If she burnt 7, how many were left?

9 Henry bought a dozen apples and gave 5 away. How many apples were left?

10 James saved $36 but spent $8. How much did he have left?

Space Pyramids

1 Shade all pyramids.

2 Tick the pyramid whose base is a square.

3 Put a cross on the pyramid with 6 corners.

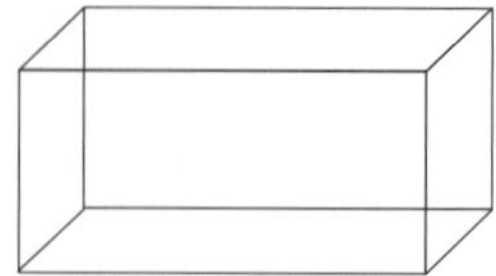
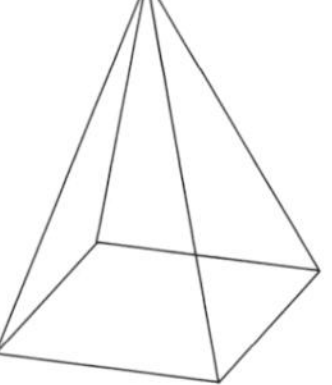

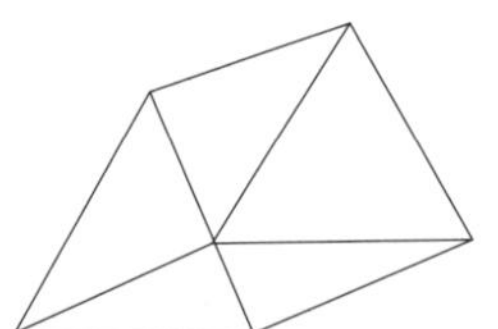
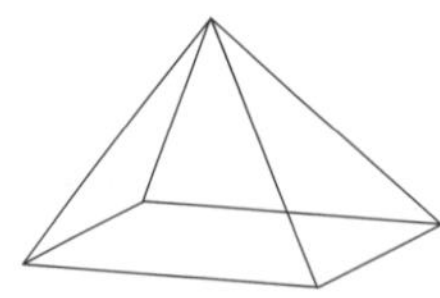
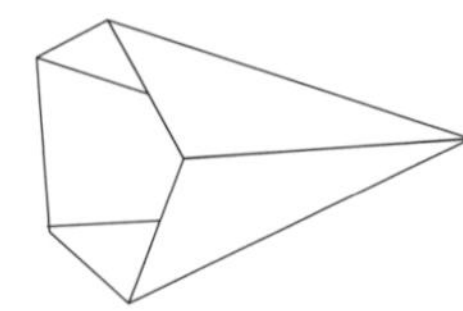
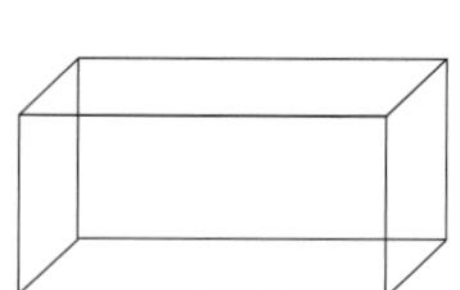
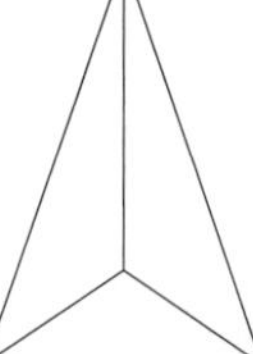

Number and Algebra

SET 3 Multiplication by 2

Use the array to help you answer the questions.

1 1 × 2 = ________
2 2 × 2 = ________
3 3 × 2 = ________
4 4 × 2 = ________
5 5 × 2 = ________
6 6 × 2 = ________
7 7 × 2 = ________
8 8 × 2 = ________
9 9 × 2 = ________
10 10 × 2 = ________

Mathematical Reasoning

11 How many different arrays can you make with 20 buttons? ________
Draw one of your arrays.

SET 4 Extension

1 What number is 3 more than 72?
2 What number is 15 less than 20?
3 What is the next odd number after 9?
4 How many days in 2 weeks?
5 Which is the greater number: 132 or 123?
6 How much change would I get from $1 if I spent 80c?
7 Share 20 counters amongst 4.
8 Ten more than seventeen
9 81 = ☐ tens + ☐ ones.
10 How many 5c coins make 50c?
11 30 + 50
12 Which number is greater, 176 or 167?
13 Write 92 in words.
14 How many eggs in 2 dozen?
15 What month comes before June?
16 Name this object.

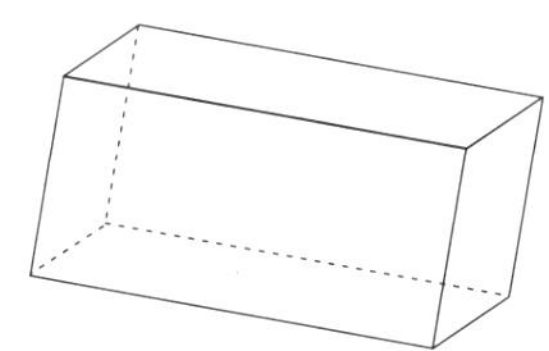

Measurement Measuring in centimetres

Estimate and then measure the length of these pencils in centimetres.

1 Estimate ☐ Actual ☐
2 Estimate ☐ Actual ☐
3 Estimate ☐ Actual ☐
4 Estimate ☐ Actual ☐
5 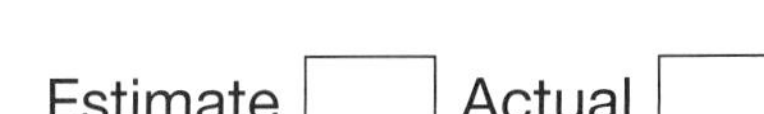 Estimate ☐ Actual ☐

UNIT 5

Number and Algebra

SET 1 Basic

1 4 + 7

2 3 + 9

3 \$7 + \$5

4 8 – ☐ = 7

5 ☐ – 9 = 1

6 3 × 4

7 7 × 2

8 6 × 4

9 ☐ – 5 = 2

10 17 + 3

11 Half of 40

12 Double 20.

13 2 lots of 4

14 5, 10, 15, ☐

15

How many oranges in 4 baskets if each basket holds 5 oranges?

☐ oranges

SET 2 Extending addition facts

1 7 + 6

2 70 + 60

3 8 + 9

4 80 + 90

5 3 + 5

6 30 + 50

7 7 + 4

8 70 + 40

9 9 + 6

10 90 + 60

11

	30	40	50	60	70	80
+	80	90	80	90	50	60

12

	20	30	40	50	600	700
+	30	40	50	60	700	900

Number and Algebra Numbers to 999

1 Write 123 in words.

☐

2 Write 367 in words.

☐

3 Write the number two hundred and thirty-seven. ☐

4 Write the number four hundred and sixty.

☐

5 Write the missing numbers.

390, 380, ☐, ☐

Write the numbers modelled on the abacuses below.

6

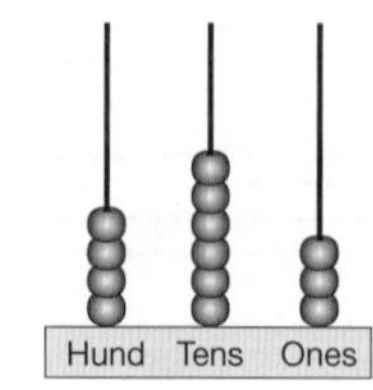

☐

7

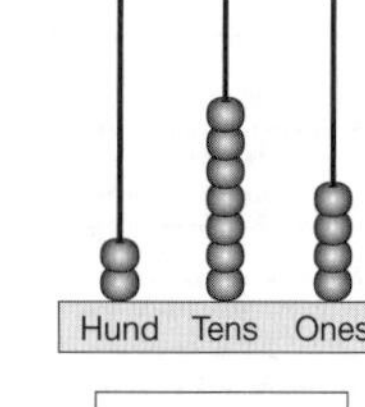

☐

Number and Algebra

SET 3 Addition patterns

Complete the addition pattern.

1

+3	
10	
15	
20	
25	
30	
35	

2

+10	
11	
21	
31	
41	
51	
61	

Mathematical Reasoning

You have $100 to spend. Use rounding skills to work out which totals are $100 or less. Tick the combinations you can afford.

$72 (necklace) $59 (watch) $41 (bangle) $28 (ring)

3 watch + ring ________

4 necklace + ring ________

5 bangle + necklace ________

6 watch + bangle ________

SET 4 Extension

1 15 + ☐ = 20

2 93 = ☐ tens and ☐ ones

3 How many rows of 5 make 20?

4 If one pencil costs 10c, how much would 4 pencils cost?

5 Write 12 in words. ☐

6 How many days in April?

7 How many months in a season?

8 What month comes after August?

9 How many eggs in $1\frac{1}{2}$ dozen?

10 Estimate an answer to 19 + 21.

11 Write the number that is 6 tens and 7 ones.

12 Jill carried 13 books but dropped 5.
She had ☐ books left.

13 Is 127 an even number?

14 Add 27 and 18.

15 How many hours from 3 pm to 9 pm?

16 Name this shape.

Space Top view

Draw a bird's-eye view of this room in the space provided.

UNIT 6

Number and Algebra

SET 1 Basic

1 7 – 6

2 9 – 5

3 9 – 0

4 1 + 6

5 4 + 5

6 0 + 9

7 2 + 10

8 Double 2.

9 10 – 4

10 9 – 6

11 One less than 15

12 Two more than 9

13 20c – 5c

14 10 – 10

15

I saved $6 in March and $14 in April. How much have I saved so far?

$ ☐

SET 2 Expanding and comparing numbers

Expand these numbers.

1 539 = 500 + ☐ + ☐

2 635 = ☐ + ☐ + ☐

3 278 = ☐ + ☐ + ☐

4 386 = ☐ + ☐ + ☐

5 794 = ☐ + ☐ + ☐

6 497 = ☐ + ☐ + ☐

Mathematical Reasoning

Use the < or > symbols to complete these sentences.

The first one is done for you.

7 400 + 20 + 8 < 600

8 600 + 80 + 5 ☐ 700

9 700 + 50 + 2 ☐ 700

10 300 + 90 + 4 ☐ 500

11 500 + 70 + 6 ☐ 500

Numbers and Algebra Numbers to 5000

Draw beads on the abacuses to represent the numbers.

1

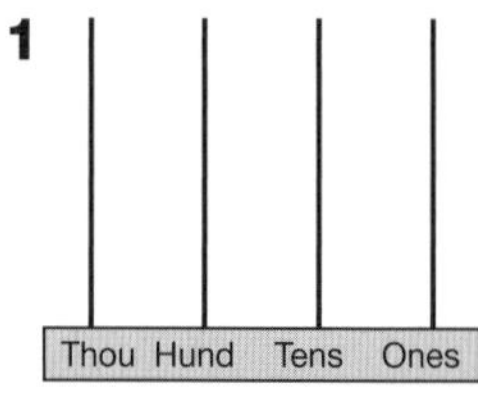

4169

2

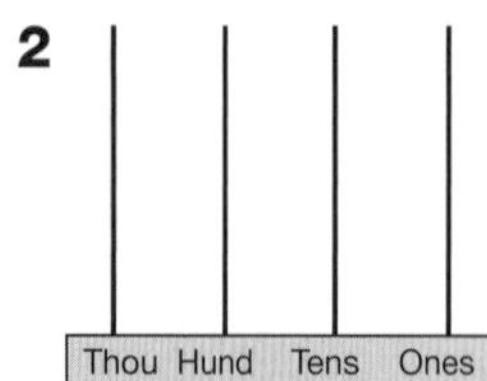

3424

3

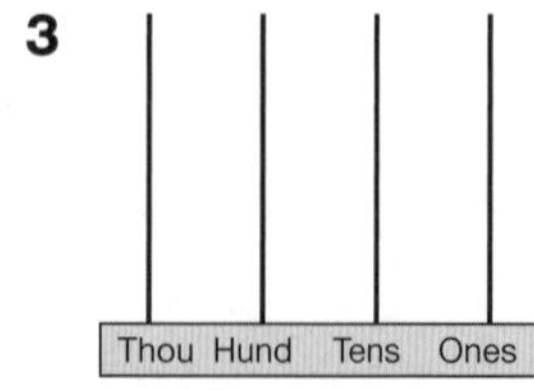

2036

4

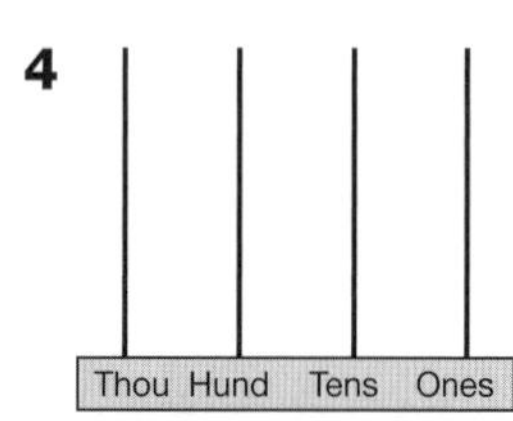

3507

Number and Algebra

SET 3 Multiplication by 5

1 [2] × 5 = 10

2 4 × [ ] = 20

3 [ ] × 5 = 20

4 [ ] × 5 = 30

5 [ ] × 5 = 15

6 7 × [ ] = 35

7 5 × 5 = [ ]

8 8 × 5 = [ ]

9 10 lots of 5 = [ ]

10 9 × 5 = [ ]

11 A farmer has 6 paddocks with 5 sheep in each. How many sheep does he have altogether? 

12 Peta earns $8 every hour she works. How much does she earn in 5 hours? 

13 Rana saves $7 each week. How much would she save in 5 weeks? 

SET 4 Extension

1 15 – 13

2 36 – 5

3 14 minus 8

4 Nine take away 4

5 15c – 10c

6 What season comes after winter?

7 How many days in April plus September?

8 Subtract 10 from two dozen.

9 Forty cents less two 10c coins

10 What number is halfway between 70 and 80?

11 4 + 41 – 2

12 150 + 30

13 Estimate an answer to 28 + 31.

14 Share 20 between 5.

Complete the grid.

	+	12	14	16	18	20	22
15	6						
16	12						

Statistics and Probability Column graphs

Use the tallied information to complete the graph.

Favourite drinks

Cola	~~IIII~~ III
Orange juice	~~IIII~~ ~~IIII~~ I
Lemonade	IIII
Lime	I
Creaming soda	~~IIII~~

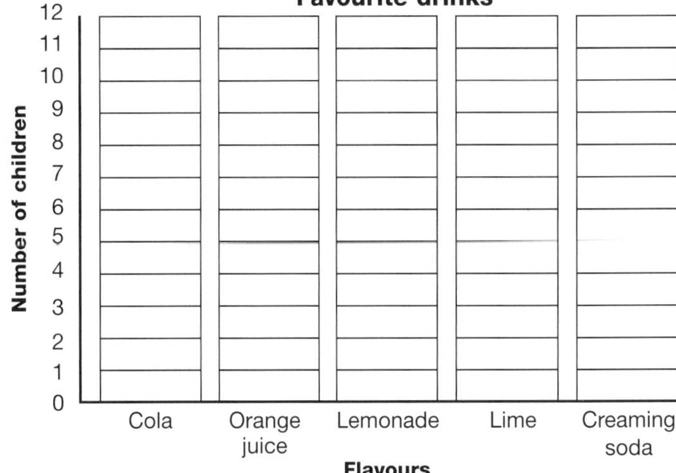

Number and Algebra

SET 1 Basic

1 7 + 2

2 5 + 4

3 2 + 3

4 5 + 5

5 8 + 2

6 8 – 4

7 7 – 4

8 9 – 6

9 Half of 12

10 Double 7.

11 2, ☐, ☐, 8, 10, ☐

12 10 + 4

13 7 + 7

14 11 – 2

15

Blake had 18 marbles but gave half away. How many marbles were left?

☐ marbles

SET 2 The division symbol

1 Share 12 fish between 4.

2 Share 12 fish between 3.

3 Share 12 fish between 2.

4 Share 12 fish between 6.

Use the apples to solve the divisions.

5 16 ÷ 2 = ☐

6 16 ÷ 4 = ☐

7 16 ÷ 1 = ☐

8 16 ÷ 8 = ☐

9 16 ÷ 16 = ☐

Numbers and Algebra Multiplication facts (10s)

1 10 lots of 3

2 10 rows of 5 plants

3 Multiply 4 by 10

4 Product of 7 and 10

5 Ten lots of zero

6 How many wheels on 10 cars?

7 How many legs on 10 octopuses?

8

	3	5	6	2	9	8	7	4
×3	9			6				

9

	3	5	6	2	9	8	7	4
×2		10		4				8

10

	3	5	6	2	9	8	7	4
×10	30		60					

Number and Algebra

SET 3 Addition/jump strategy

1 Add 15 and 4.

2 10 + 15

3 26 plus 5

4 What is the sum of 17 and 7?

5 What is 5 more than 33?

6 What is 6 more than 18?

7 What is 39 plus 7?

8 93 + 0

9 49 + ☐ = 55

10 Add $7 to $13.

11 11 cm + 20 cm = ☐ cm

12 77 plus 8

Use the number line to solve these.

13 22 + 13 =

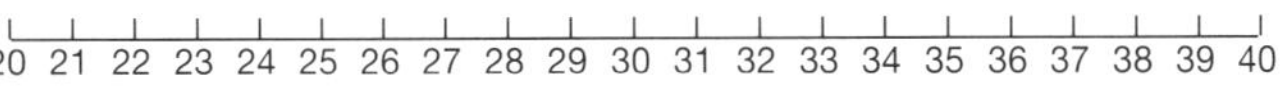

14 29 + 11 =

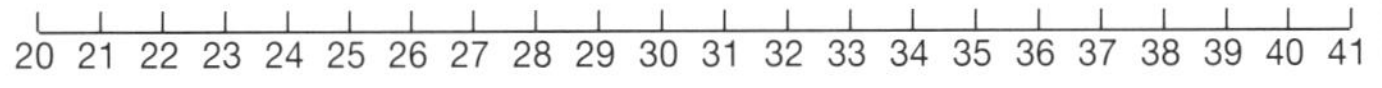

SET 4 Extension

1 Share $15 among 3.

2 What is the next odd number after 79?

3 Which is the larger number, 64 or 46?

4 How many 5c coins make 80c?

5 Write 16 in words. ☐

6 Is 39 an even number?

7 What number comes 3 before 72?

8 How many girls are there if there are 6 rows of 4 girls?

9 How many 20c coins in $1.80?

10 What is the fifth month of the year?

Mathematical Reasoning

11 Place the numbers 2, 4 and 7 in the circles so that the sum of each line is 15.

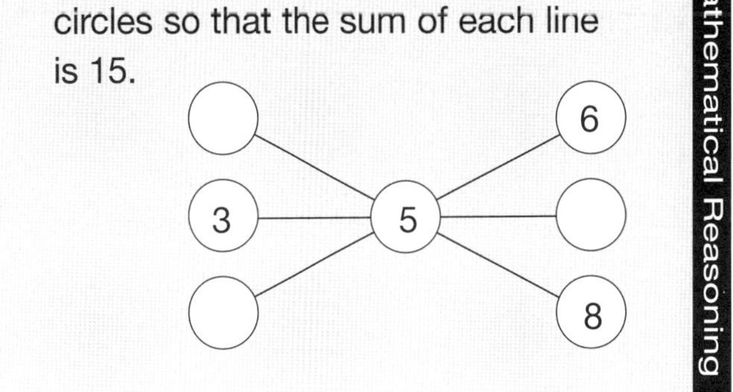

Measurement Informal capacity

Circle the most suitable container to measure the capacity of the bathtub and the saucepan.

1

2

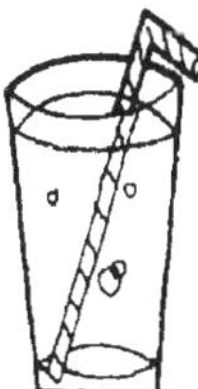

UNIT 8

Number and Algebra

SET 1 Basic

1 3 + 3

2 4 + 4

3 8 + 3

4 6 + 3

5 19 + ☐ = 20

6 9 – 5

7 9 – 3

8 9 – 2

9 9 – 9

10 7 × 4

11 8 × 4

12 10 × 2

13 Half of 100

14 Sum of 4 and 7

15 

SET 2 Extending subtraction facts

1 7 – 2

2 70 – 20

3 9 – 5

4 90 – 50

5 13 – 6

6 130 – 60

7 8 – 4

8 80 – 40

9 800 – 400

10 500 – 300

11 Barry's mass is 70 kg and Jill's mass is 40 kg. What is the difference in their masses? 

12 Hannah needed $800 to buy a new washing machine. If she has saved $300, how much more does she need to save? 

Number and Algebra Bar models/addition

Use the bar models to help you solve the additions.

1 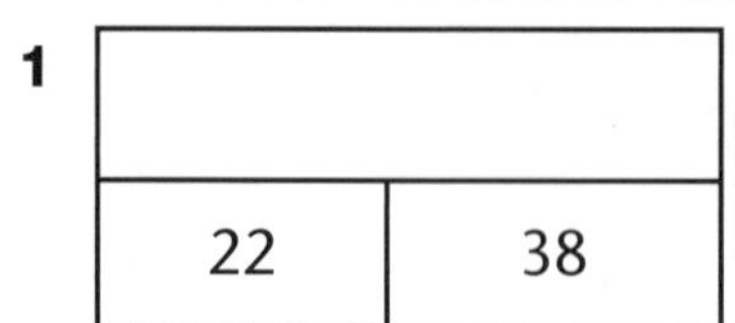

22 + 38 = ☐

4 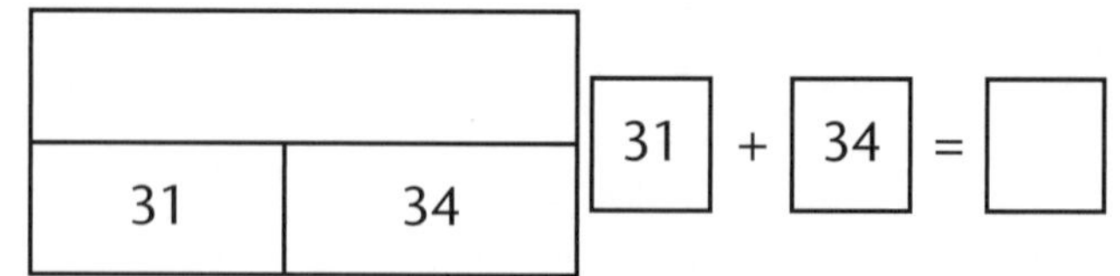

2 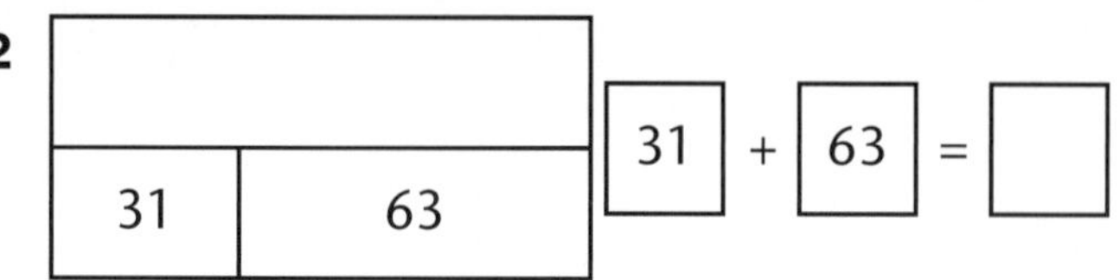

5 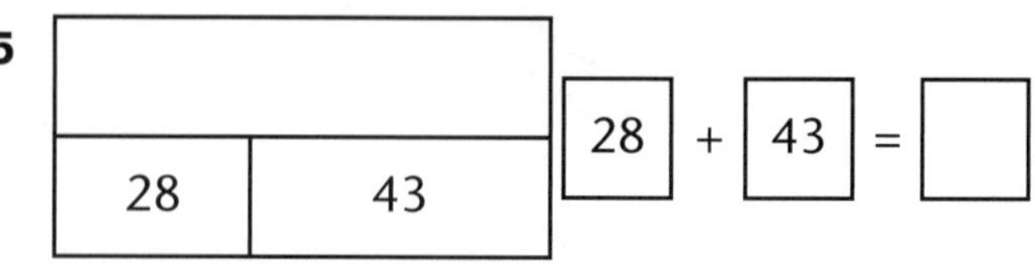

3 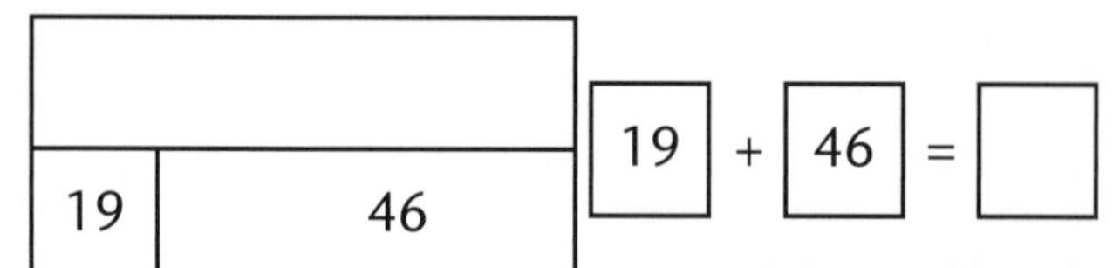

6 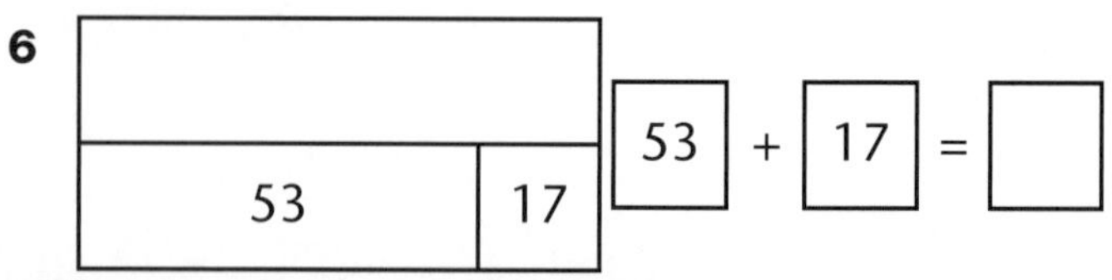

Number and Algebra

SET 3 Unit fractions

Colour the fraction of each pizza.

1 $\frac{1}{8}$

2 

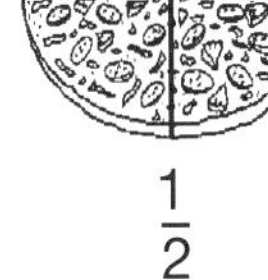

$\frac{1}{2}$

3 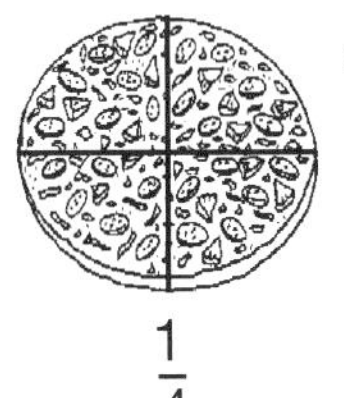

$\frac{1}{4}$

4 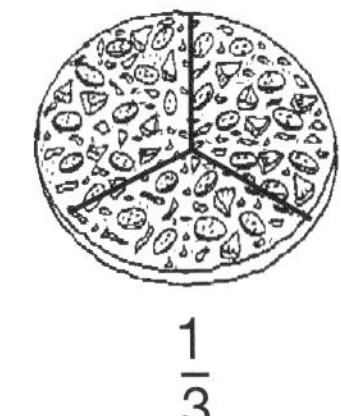

$\frac{1}{3}$

5 Order the fractions from smallest to largest by numbering them 1 to 4.

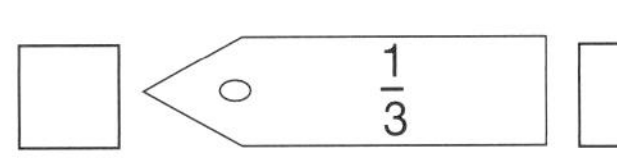

$\frac{1}{2}$ ☐ $\frac{1}{3}$ ☐

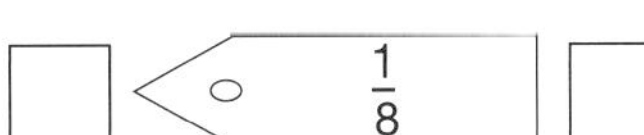

$\frac{1}{4}$ ☐ $\frac{1}{8}$ ☐

6 Circle the groups into the given fractions.

a $\frac{1}{2}$

b $\frac{1}{4}$

SET 4 Extension

1 4 + 4 + 4

2 15 – 0

3 Share 18 among 3.

4 Which is the third month of the year?

5 130 + 60

6 Write 17 in words.

7 How many wheels are there on 7 push bikes?

8 Subtract 5 from 91.

9 What is the sum of 17 and 24?

10 What is the sum of 19 and 12?

11 How many days are in June?

12 How many halves are in a whole?

13 Estimate an answer to 42 + 29.

Mathematical Reasoning

14 Write 4 number sentences that equal 27.

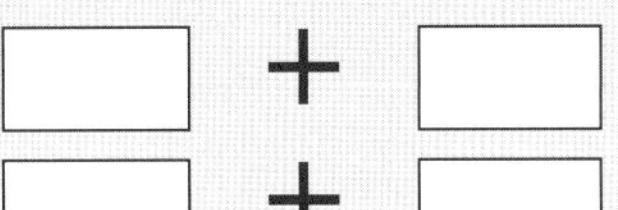

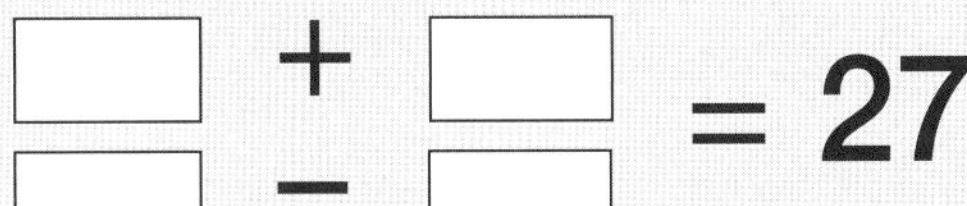

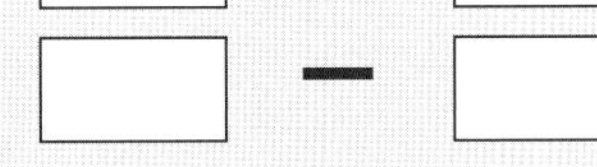

☐ + ☐
☐ + ☐
☐ – ☐
☐ – ☐
= 27

Space Surfaces of 3D objects

List the shapes that could be used to make each object. (E.g., squares)

1 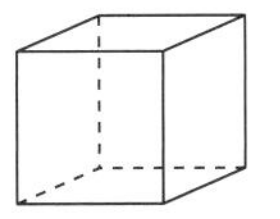

square prism

2 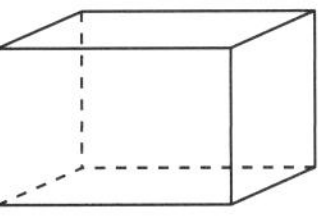

rectangular prism

3 

triangular prism

4 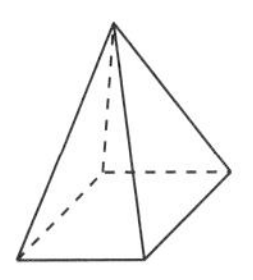

square pyramid

UNIT 9

Number and Algebra

SET 1 Basic

1 8 – 5

2 10 – 4

3 10 – 1

4 2 + 7

5 3 + 6

6 1 + 8

7 3 + 9

8 Double 9.

9 10c – 5c

10 8 – 7

11 Two less than 17

12 Two more than 8

13 6 × 2

14 6 × 4

15

John has 6 popsticks in one hand and 9 in his other hand. How many popsticks is John holding altogether?

☐ popsticks

SET 2 Split strategy for addition

Add the tens then the ones.
24 + 38 becomes 50 + 12 = 62.

Add the following numbers by adding the tens part then the ones part.

1 27 + 32 becomes ☐ + ☐ = ☐

2 44 + 23 becomes ☐ + ☐ = ☐

3 35 + 17 becomes ☐ + ☐ = ☐

4 58 + 22 becomes ☐ + ☐ = ☐

Complete the grid to show the sum of each pair of numbers.

	Numbers	Sum
5	29 and 16	45
6	10 and 23	
7	42 and 17	
8	16 and 32	
9	36 and 15	
10	45 and 23	

Statistics Picture graphs

Mr Black's class hair colours

Blond	5 faces
Red	2 faces
Brown	9 faces
Green	1 face
Black	5 faces
Fair	7 faces

Use the graph to answer the questions.

1 How many children have blond hair? ☐

2 How many children have brown hair? ☐

3 How many children have red hair? ☐

4 How many children are in Mr Black's class? ☐

5 Which colours have the same number of children?

___

6 Write a question for this graph.

___

Number and Algebra

SET 3 Trading in addition

Complete the additions.

1

	Tens	Ones
	5	7
+	2	6

2

	Tens	Ones
	4	9
+	2	5

3

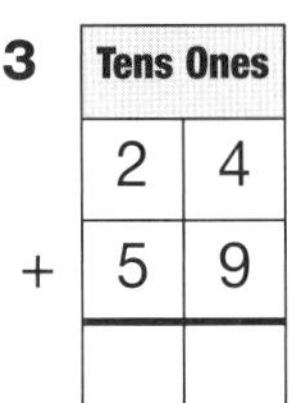

	Tens	Ones
	2	4
+	5	9

4

	Tens	Ones
	3	6
+	4	6

5

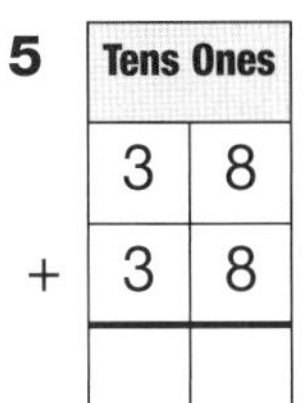

	Tens	Ones
	3	8
+	3	8

6

	Tens	Ones
	4	6
+	4	7

7

	Tens	Ones
	4	8
+	3	7

8

	Tens	Ones
	3	5
+	3	9

SET 4 Extension

1. 4 × ☐ = 8
2. How many trees are there, if there are 7 rows of 4 trees?
3. How many books are there if there are 10 bundles of 8 books?
4. What number is 7 more than 23?
5. Write 23 in words. ☐
6. Share 16 among 4.
7. Which month has the least days?
8. How many sides on 6 rectangles?
9. What number comes 3 before 98?
10. Share 25 among 5.
11. 85 = ☐ tens and ☐ ones.
12. How many halves in 2 wholes?
13. What is the eighth month of the year?
14. 10 pairs of socks = ☐ socks
15. Shade 4 numbers that add to make 47.

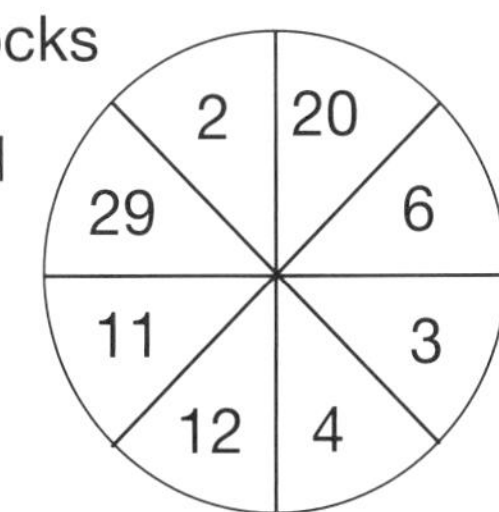

Measurement Quarter to and quarter past

Draw the times on the clock faces.

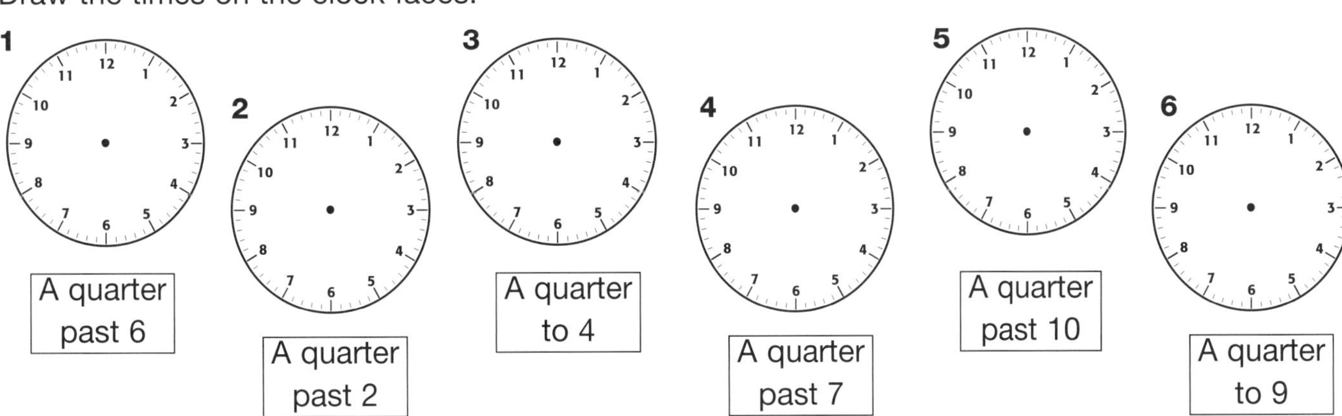

Number and Algebra

SET 1 Basic

1 9 + 4

2 \$9 – \$7

3 3 × 4

4 10 + 2

5 12 – 2

6 6 × 4

7 4 × 4

8 9 take away 4

9 4 less than 17

10 Double 8.

11 90c + 5c

12 16 minus 10

13 20 plus 10

14 2 × 10

15

Trent saved \$6 one week and \$8 the next week. What was his total savings for the 2 weeks?

\$ ☐

SET 2 Subtraction using addition

Solve each subtraction fact then write an addition fact to check your answer.

	Subtraction	Addition
1	17 – 6 = 11	6 + 11 = 17
2	23 – 9 = ☐	☐ + ☐ = ☐
3	35 – 31 = ☐	☐ + ☐ = ☐
4	57 – 52 = ☐	☐ + ☐ = ☐
5	27 – 18 = ☐	☐ + ☐ = ☐
6	45 – 32 = ☐	☐ + ☐ = ☐

7

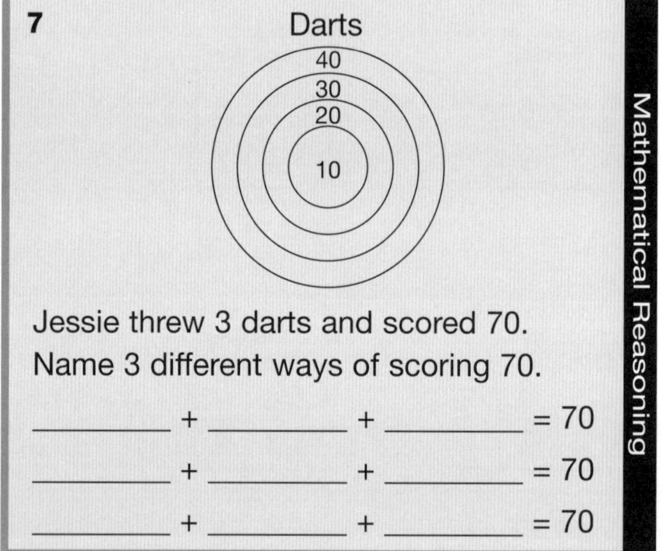

Jessie threw 3 darts and scored 70.
Name 3 different ways of scoring 70.

_______ + _______ + _______ = 70

_______ + _______ + _______ = 70

_______ + _______ + _______ = 70

Space Angles

Colour all the angles blue that are smaller than a right angle. Colour all the angles red that are larger than a right angle.

1 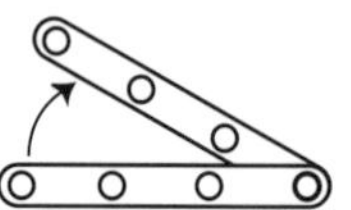

2 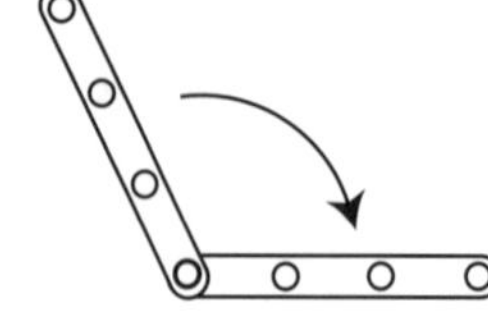

3 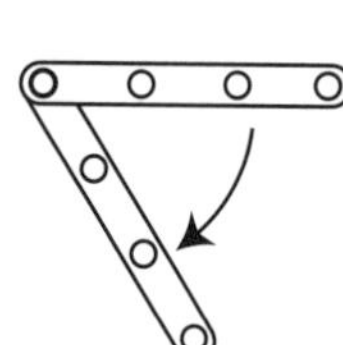

4 

5 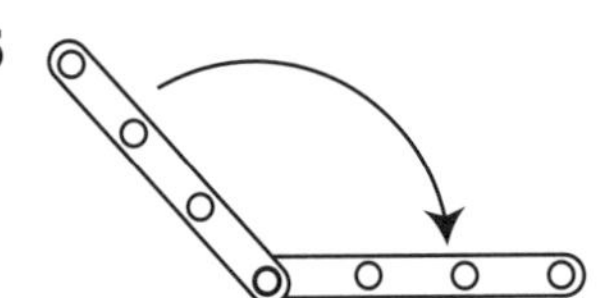

Number and Algebra

SET 3 Number patterns

Complete each pattern then write the rule for it. Check each pattern by backtracking.

1

4	6	8	10				

Rule: ____________

2

3	6	9	12				

Rule: ____________

3

15	20	25	30				

Rule: ____________

Mathematical Reasoning

4 Draw the 3rd and 5th staircase numbers in the pattern.

5 What's the next number in the pattern? ________

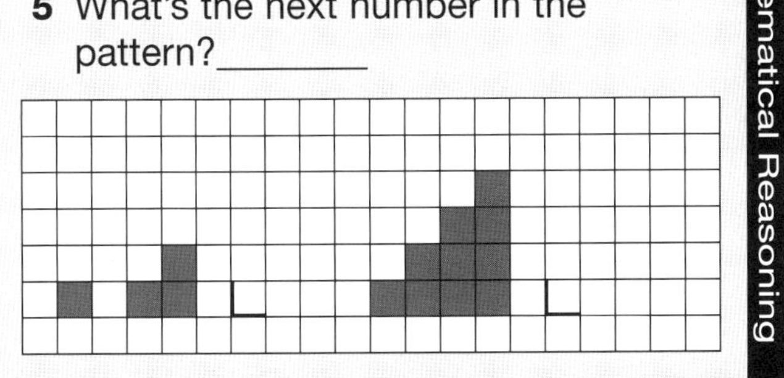

SET 4 Extension

1 82 = ☐ tens + ☐ ones

2 Which is larger: 6 + 4 or 3 + 8?

3 Twenty-one and nine more

4 35c + 55c

5 How many sides on 6 triangles?

6 If Wednesday is 1 February, what is Thursday's date?

7 10 + 10 + 10 + 5

8 How many $5 notes in $50?

9 Share 18 among 7.

10 Write 95 in words. ____________

11 Forty-six and sixteen more

12 234 = ☐ hund + ☐ tens + ☐ ones

13 35 – 17

14 Write 28 in words. ____________

15 Mt Cook Primary School has 37 boys and 56 girls. How many children attend the school altogether? 

Measurement The metre

Colour the correct box to record the heights of the following items.

	Item	Taller than 1 m	About the same as 1 m	Shorter than 1 m
1	broom			
2	chair			
3	refrigerator			
4	dustbin			
5	shovel			
6	telephone booth			

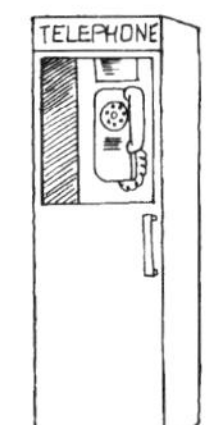

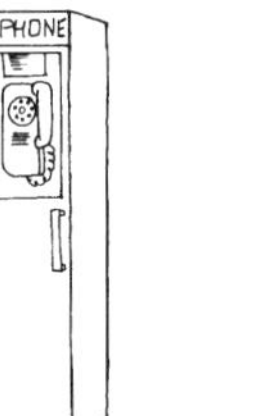
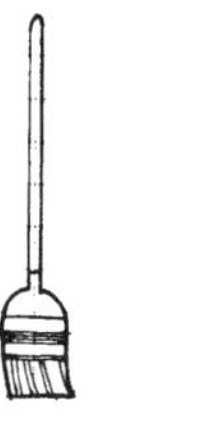
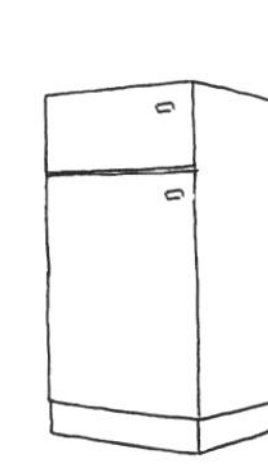

Number and Algebra

SET 1 Basic

1 5 + 5

2 6 + 4

3 14 + 6

4 19 + 1

5 10 – 7

6 10 – 6

7 10 – 5

8 10 – 4

9 4 × 3

10 4 × 6

11 4 × 4

12 Sum of 11 and 9

13 6 plus 8

14 Six more than 9

15 

Jill had $17 but Jack had only $5. How much more money did Jill have?

$ ☐

SET 2 Counting forwards and backwards

1 Count forward by tens.

14	24						

2 Count forward by tens.

130	140						

3 Count forward by tens.

57	67						

4 Count backwards by tens.

93	83						

5 Count backwards by tens.

100	90						

6 Count backwards by tens.

255	245						

7 34, 44, 54, ☐, ☐

8 122, 132, 142, ☐, ☐

9 54, 64, 74, ☐, ☐

10 125, 135, 145, ☐, ☐

11 150, 160, 170, ☐, ☐

Number and Algebra Bar models/subtraction

Use the bar models to help you complete the subtractions.

1 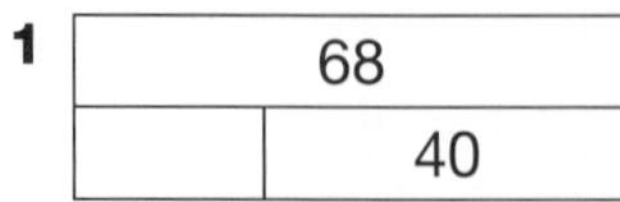

68 – 40 = ☐

2 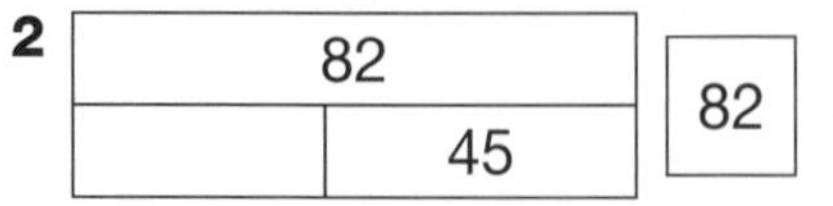

82 – 45 = ☐

3 65 / 39

65 – 39 = ☐

4 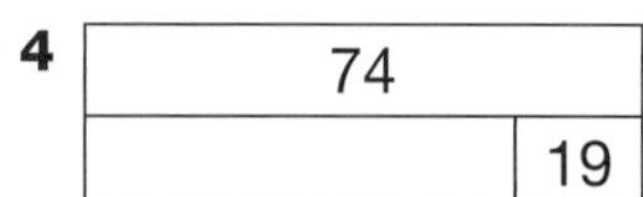

74 – 19 = ☐

5 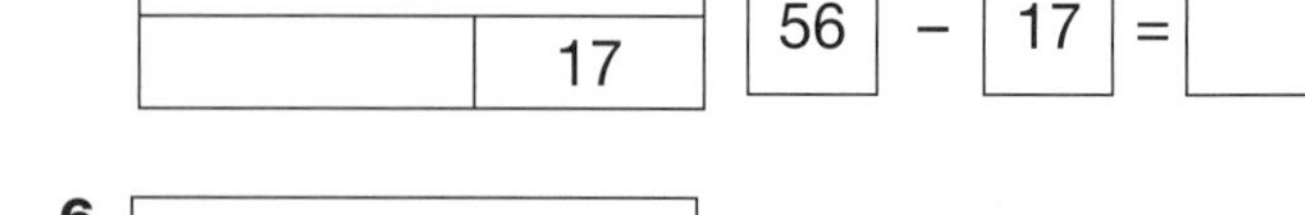

56 – 17 = ☐

6 43 / 28

43 – 28 = ☐

Number and Algebra

SET 3 Halves and quarters

Shade the given fraction of each shape or group.

1 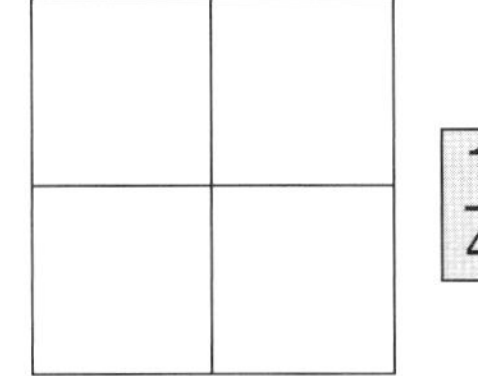 $\frac{1}{4}$

4 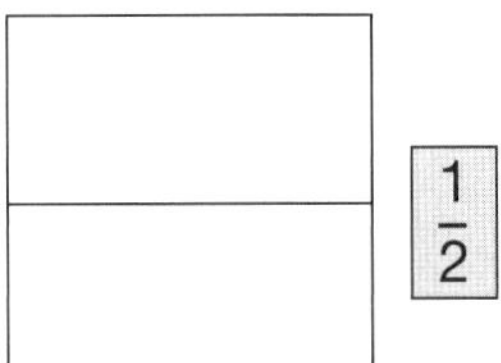 $\frac{1}{2}$

2 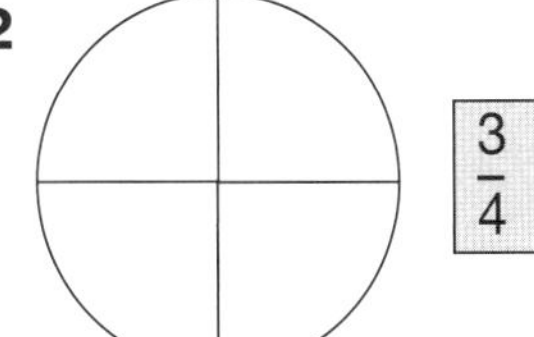 $\frac{3}{4}$

5 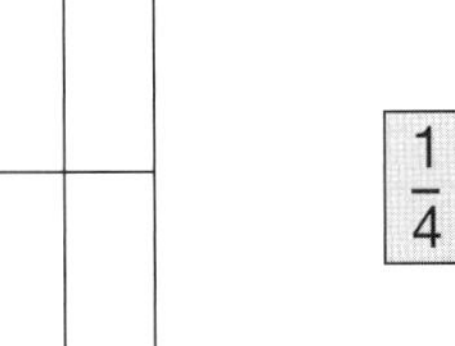 $\frac{1}{4}$

3 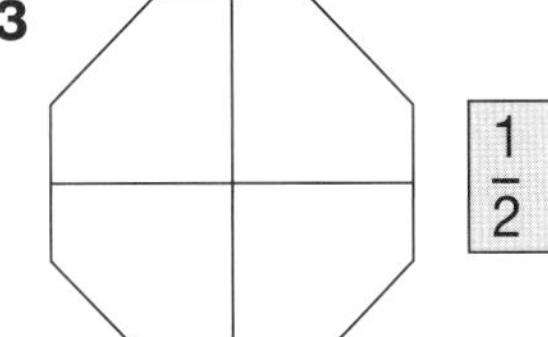 $\frac{1}{2}$

6 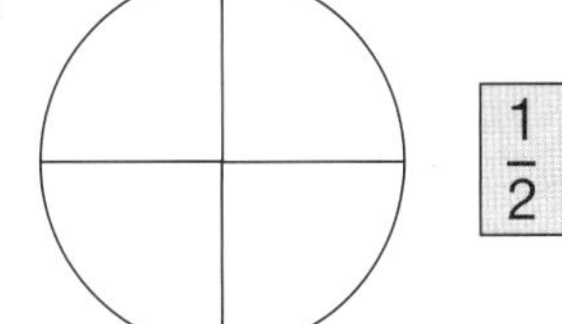 $\frac{1}{2}$

7 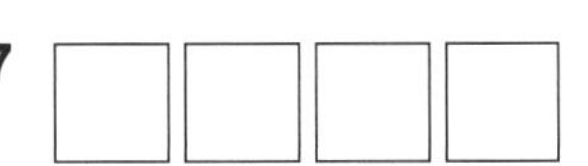 $\frac{3}{4}$

8 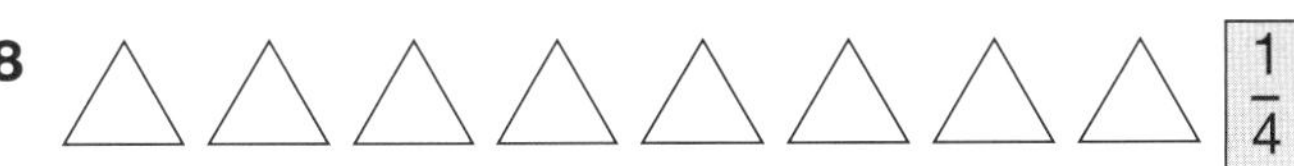 $\frac{1}{4}$

9 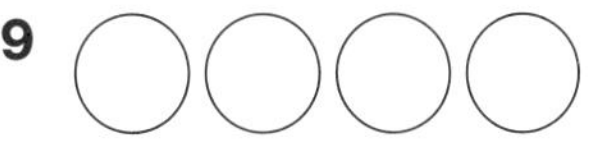 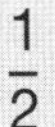 $\frac{1}{2}$

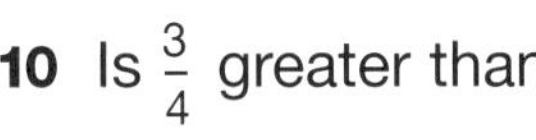
10 Is $\frac{3}{4}$ greater than $\frac{1}{2}$?

SET 4 Extension

1 Half of 32

2 Put these numbers in counting order:

17, 12, 11, 19.

3 23 + 35

4 Write 187 in words.

5 3 hundreds + 2 tens + 5 ones

6 Cents in $1.75

7 How many 10c coins in $1.50?

8 How much are 10 cakes at 15c each?

9 How many sides on 7 triangles?

10 How many centimetres in 2 m?

11 Twelve less than 58

12 Jane had $2 and spent $1.50. How much was left?

13 27 + ☐ = 40

14 36 = ☐ dozen

15 How many hours are between 11 am and 4 pm?

16 Name this shape. 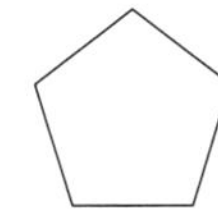

Statistics Picture graphs

Favourite fruits

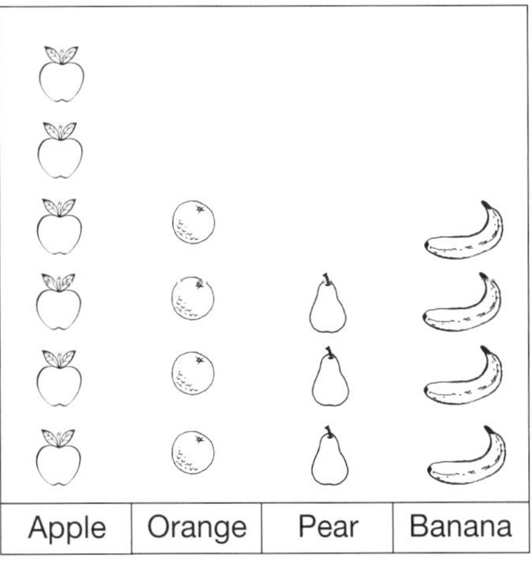

Use the key and the graph to answer the questions.

1 Which fruit was most popular?

2 Which fruit scored 15 on the graph?

3 Which fruits were equally popular?

4 Which fruit scored 10 more than bananas?

KEY	 1 fruit = 5 pieces

Number and Algebra

SET 1 Basic

1 20 + 0

2 14 + 3

3 12 + 6

4 9 + 7

5 8 + 5

6 7 + 4

7 15 + 2

8 17 + 1

9 9 × 2

10 9 × 4

11 20 – 10

12 17 – 3

13 18 – 14

14 7 less than 17

15

Justin has $8. If Kirsty has $7 more, how much does Kirsty have?

$ ☐

SET 2 Doubles and near doubles

1

Double							
0	1	2	9	4	8	6	12

2

Double							
10	40	30	70	60	20	22	80

Use doubles or near doubles to answer the questions.

3 20 + 19 =

4 40 + 42 =

5 30 + 31 =

6 22 + 31 =

Mathematical Reasoning

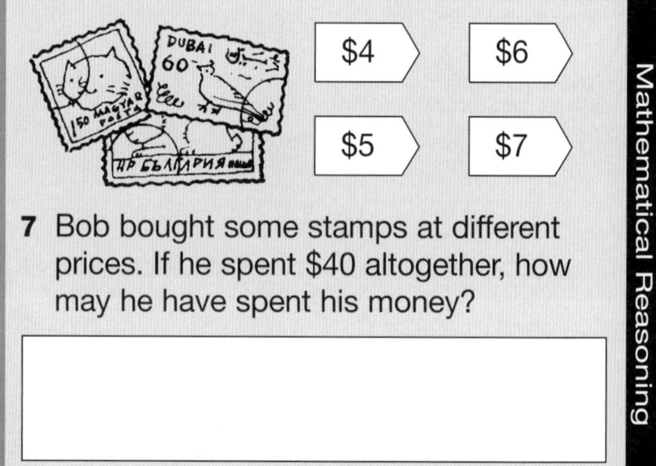

7 Bob bought some stamps at different prices. If he spent $40 altogether, how may he have spent his money?

Space Symmetry

Use the line of symmetry to complete the shapes.

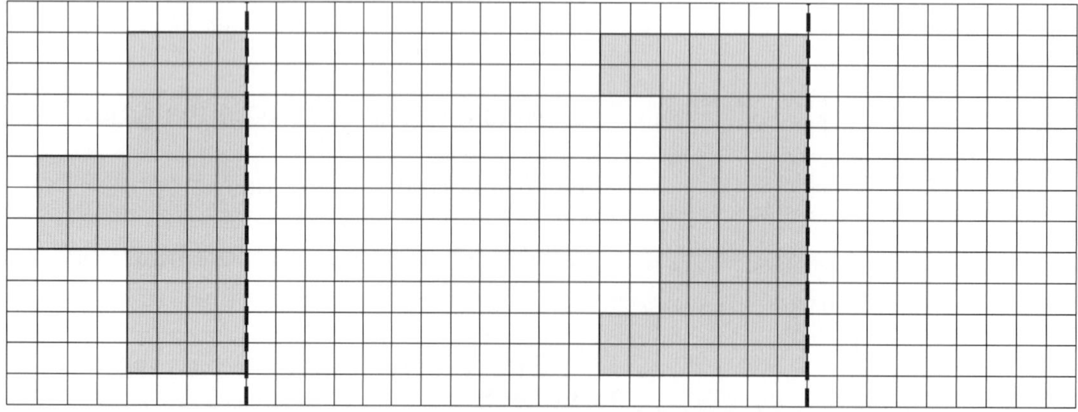

Number and Algebra

SET 3 Doubling and halving

Continue the doubling patterns.

1

2	4				

2

3	6				

3

5	10				

Continue the halving patterns.

4

128	64	32			

5

544	272	136			

6

608	304	152			

7

448	224	112			

SET 4 Extension

1 What is the hottest season?

2 Put these numbers in counting order: 7, 3, 2, 5, 1.

3 $\frac{1}{2}$ of 26

4 How many tens in 96?

5 45 + 13

6 Write 64 in words.

7 8 tens and 9 ones

8 Share 36 among 3.

9 How many 20c coins in $3?

10 How much are 1 dozen chocolates at $2 each?

11 How many wheels on 7 cars?

12 How many centimetres in a metre?

13 Cents in $1.20

14 Fifteen less than 40

15 There are 35 children in Yr 3, 27 in Yr 4, 20 in Yr 5 and 32 in Yr 6. How many children are there altogether?

Statistics Chance

Sally had red shorts and blue shorts. She also had a red top and a blue top. Colour the clothes below to find the four combinations of colours that she could wear.

Number and Algebra

SET 1 Basic

1 8 + 1 + 0

2 8 + 1 + 2

3 4 × 4

4 $10 – $6

5 12 + 7

6 13 + 5

7 20 + 0

8 7 × 3

9 18 – 5

10 25 – 20

11 19 minus 7

12 Ten less than twenty-four

13 Double 10.

14 Half of 12

15 

There are 25 pages in the book. Jill has read 17. How many pages are there left to read?

☐ pages

SET 2 Trading in subtraction

Complete the subtractions.

1

	TENS	ONES
	7	8
–	2	3

2

	TENS	ONES
	8	9
–	3	4

3

	TENS	ONES
	9	2
–	3	6

4

	TENS	ONES
	7	3
–	4	5

5

	TENS	ONES
	8	3
–	2	5

6

	TENS	ONES
	7	1
–	2	3

7

	TENS	ONES
	6	3
–	2	5

8

	TENS	ONES
	8	2
–	2	4

Space Angles

1 Draw an angle smaller than a right angle.

2 Draw an angle larger than a right angle.

3 Colour the grid to classify the angles.

Angle	Smaller than a right angle	Right angle	Larger than a right angle
(acute angle)			
(right angle)			
(obtuse angle)			

Number and Algebra

SET 3 Thirds, halves and quarters

Shade the given fraction of each shape.

1 $\frac{1}{4}$

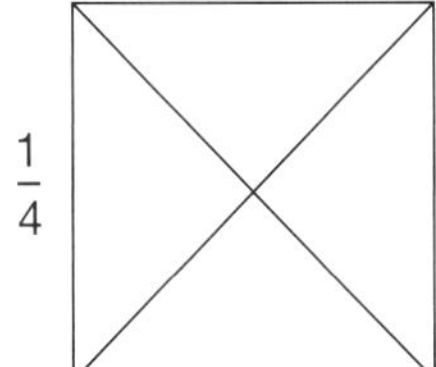

2 $\frac{1}{3}$

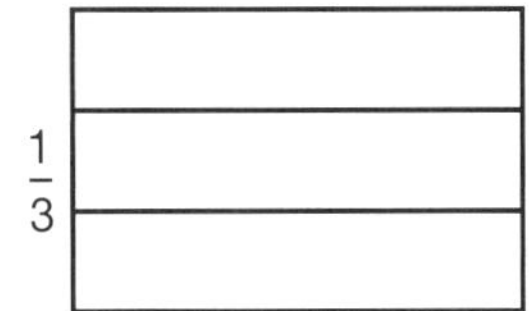

3 $\frac{1}{2}$

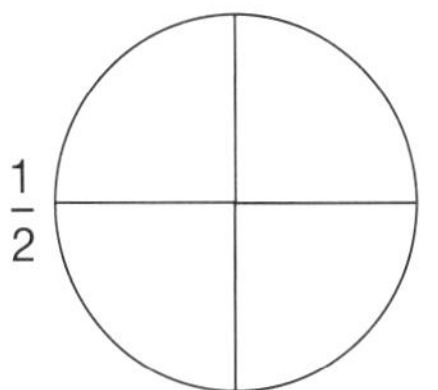

4 $\frac{1}{3}$

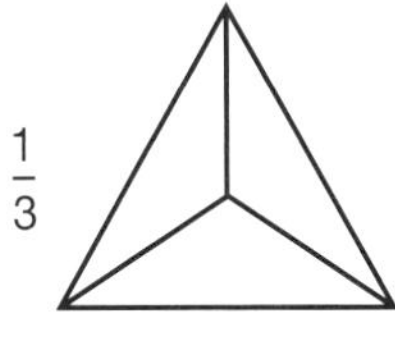

Shade $\frac{1}{3}$ of each group.

5

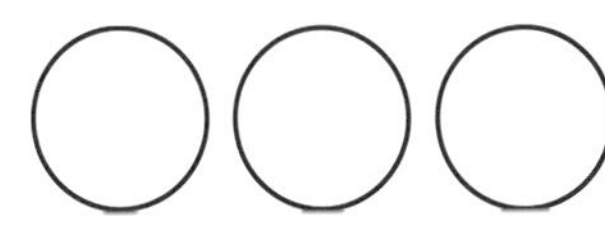

6

7

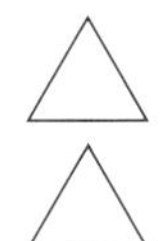

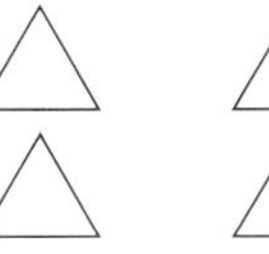
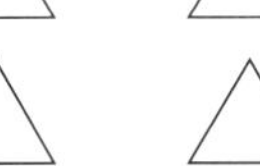

SET 4 Extension

1 Fifty-two and thirty more

2 7 tens + 3 ones

3 How many sides has a triangle?

4 How many sides in 5 triangles?

5 \$28 – \$9

6 What is the number 20 more than 17?

7 52 – ☐ = 30

8 Three dozen minus twelve

9 Share 15 between 5.

10 How many 10c coins in \$2?

11 Which month comes before September?

12 Change from \$1 after spending 35c

13 Hours in $2\frac{1}{2}$ days

14 What is the product of 4 and 5?

15 If this month is June, what month will it be in 3 months' time?

16 Colour the third bear from the left.

Measurement Litres

Find the differences in capacity between the following:

1 the bucket and the ice-cream container

2 the cordial flask and the milk carton

3 the bucket and 2 full cordial flasks

4 the oil can and 4 lemonade bottles

5 How many times must an ice-cream container be filled in order to fill the bucket?

6 How many full cordial flasks can be poured into the bucket?

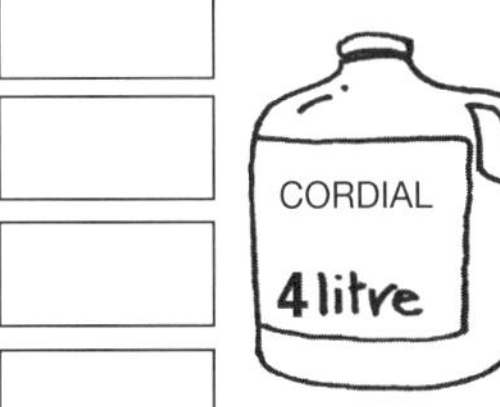

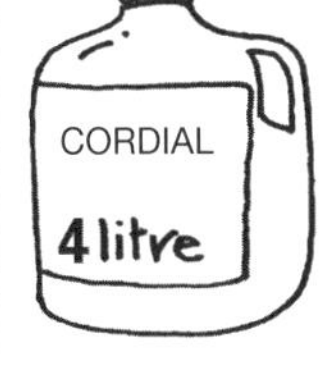

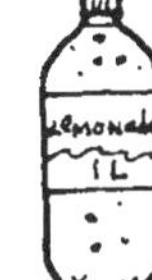

Number and Algebra

SET 1 Basic

1 6 + 3 + 1

2 7 + 2 + 2

3 18 – 4

4 4 + 2 + 3

5 Two less than 7

6 5 + 2 + 5

7 Double 11.

8 8 + 5 + 4

9 10 + 2 + 3

10 One more than 18

11 9 + 3 + 2

12 Is 21 an odd number?

13 2 + 1 + 4

14 5 × 4

15

SET 2 Division using arrays

Use the stars to help solve the divisions.

1 24 ÷ 4 = ☐

2 24 ÷ 6 = ☐

3 24 ÷ 2 = ☐

4 24 ÷ 12 = ☐

5 24 ÷ 8 = ☐

6 24 ÷ 3 = ☐

Answer these questions.

7 15 ÷ 3 = ☐

8 15 ÷ 5 = ☐

9 20 ÷ 4 = ☐

10 25 ÷ 5 = ☐

11 30 ÷ 6 = ☐

Space Modelling objects/nets

1 Colour the nets that would fold to make a cube.

a

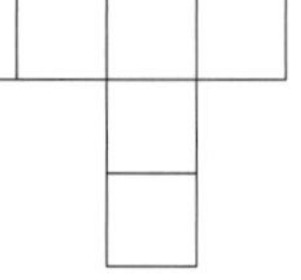

b

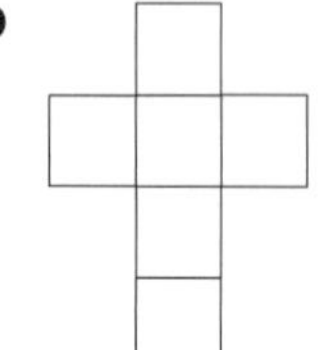

c

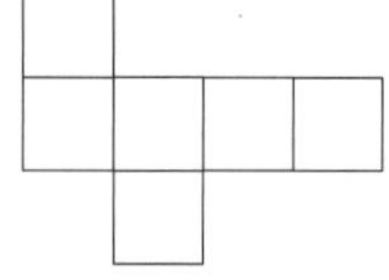

d

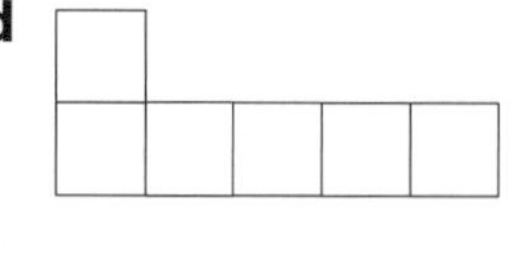

2 Count the number of cubes in each object.

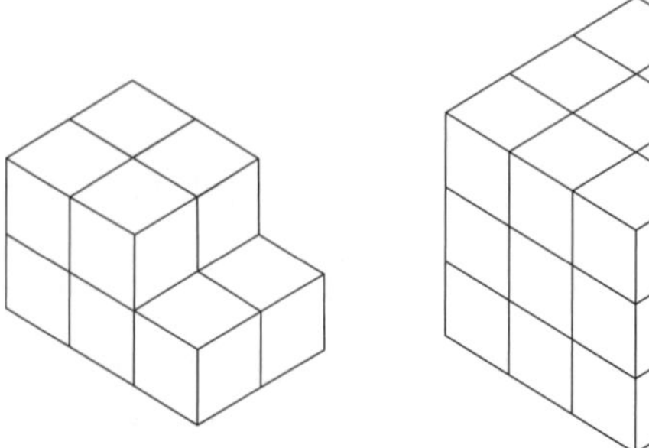

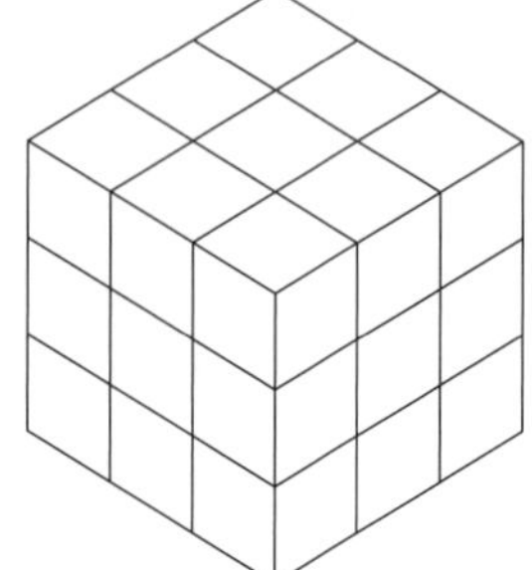

☐ Cubes ☐ Cubes

Number and Algebra

SET 3 Rounding numbers

Round these numbers to the nearest 10 before adding them to make an estimate.

1 39 + 21 = ☐

2 58 + 19 = ☐

3 43 + 49 = ☐

4 38 + 42 = ☐

5 51 + 37 = ☐

6 69 + 22 = ☐

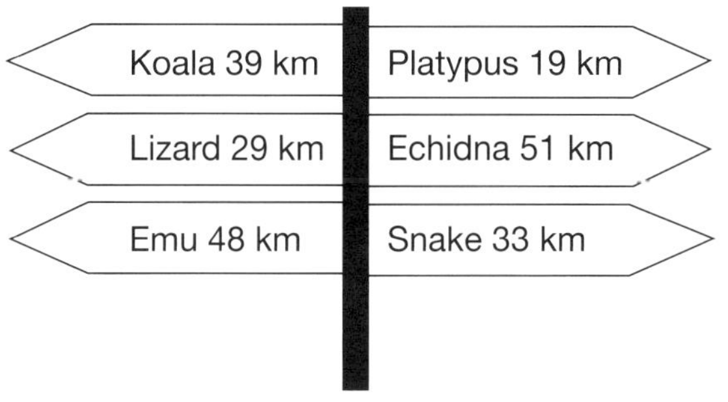

Estimate the distances from:

7	Koala to Platypus	km
8	Lizard to Echidna	km
9	Emu to Snake	km

SET 4 Extension

1 Add 7 to 206.

2 How many tens in 300?

3 Write four hundred and six.

4 How many $2 coins make $50?

5 37 – ☐ = 12

6 56 + ☐ = 80

7 276 = ☐ hund + ☐ tens + ☐ ones

Mathematical Reasoning

Fill in the empty boxes to create number sentences that are true.

8 ☐ + ☐ = 16

9 ☐ + ☐ > 25

10 ☐ × ☐ > 23

11 ☐ ÷ ☐ = 4

Colour the two parts of each circle that multiply to give 28.

12

13

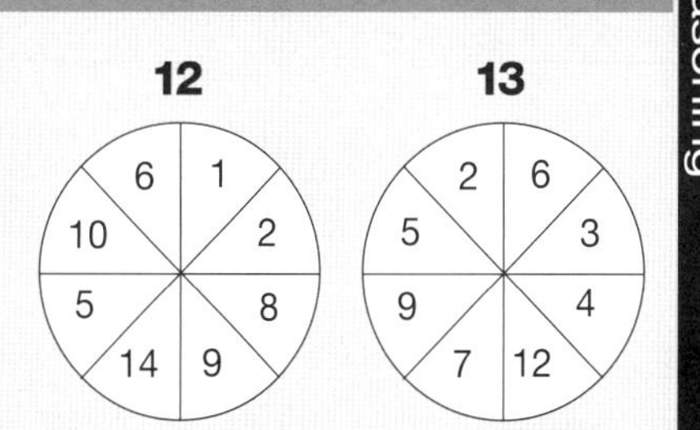

Statistics Data displays/tables

Likes rugby	Likes soccer
Kim	Jessica
John	Kelly
Tim	Con
Greg	Abdul
Kelly	Tim
Maria	Sam
Soula	John

1 Which children like both rugby and soccer?

____

2 Which children like soccer only?

____

____

3 Which children like rugby only?

____

____

UNIT 15

Number and Algebra

SET 1 Basic

1 2 + 11

2 3 + 12

3 5 + 9

4 16 + 7

5 12 – 2

6 11 – 2

7 10 – 2

8 19 – 7

9 What is the sum of 12 and 8?

10 7 × 2

11 3 × 4

12 What is the product of 3 and 4?

13 Write the number that follows 324.

14 What is the sum of 3 and 10?

15

SET 2 Money

SPECIAL $1.85

Draw strokes in the grid to show 6 different ways you could pay for a lolly bag. An example has been given.

	5	10	20	50	1
	I	II	III	II	
1					
2					
3					
4					
5					
6					

7 Subtract 20c from $2.00.

8 Ten lots of 20c

9 I had $29 but lost $12. How much have I now?

10 How many 10c coins in $2.20?

Space Top view

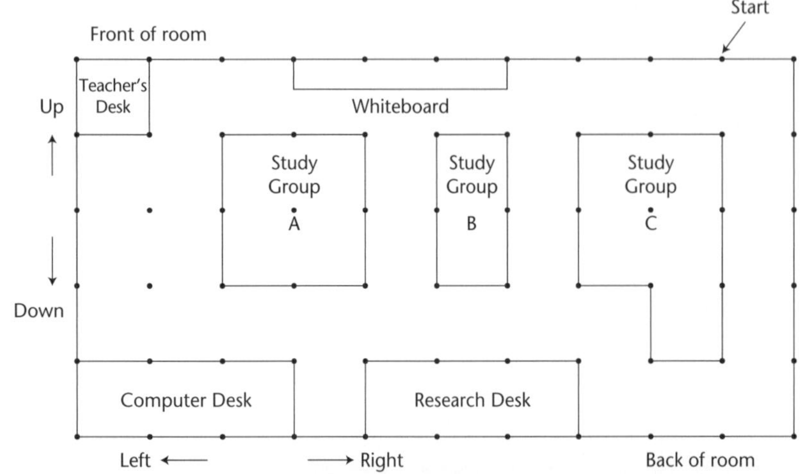

Follow the directions:

1 Starting at the starting point, move down one space.

2 Move left three spaces.

3 Move down three spaces.

4 Move left five spaces.

5 Move up 3.

6 What did you find? ___________

Number and Algebra

SET 3 Numbers to 10 000

Expand the numbers on the numeral expanders.

1 3674

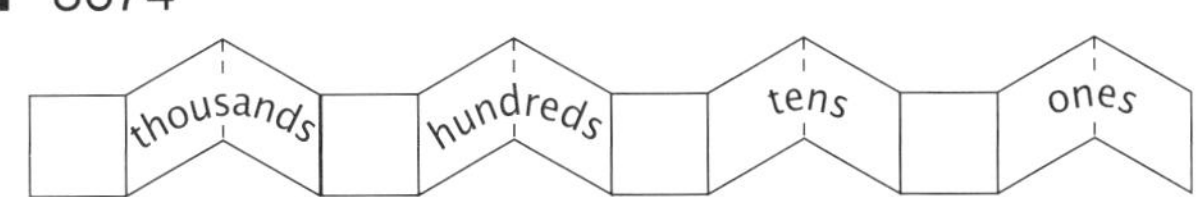

2 6871

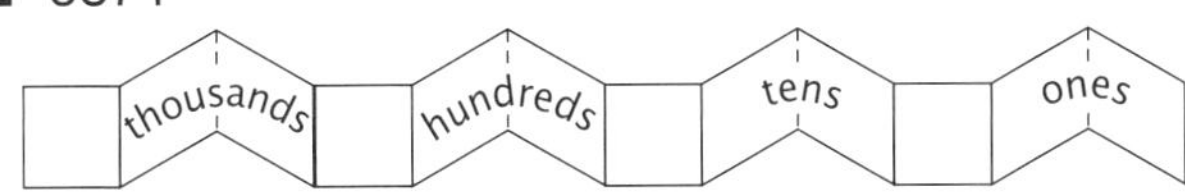

3 8150

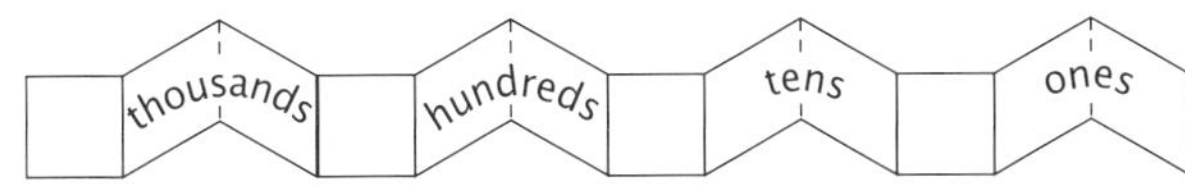

Record the place value of each bold number.

4 8**7**43 ____________

5 65**7**9 ____________

6 5**5**49 ____________

7 **3**567 ____________

8 970**9** ____________

9 **3**474 ____________

SET 4 Extension

1 Add 14 to 31.

2 What is the product of 4 and 6?

3 How many days in 3 weeks?

4 Estimate an answer to 37 + 69.

5 Round 363 to the nearest 10.

6 65 + ☐ = 80

7 Total of 13 and 12

8 3 hundreds + 2 tens + 7 ones

9 Share 40 among 5.

10 How many centimetres in 3 m?

11 Write 172 in words.

☐

12 Take 45 from 85.

13 48 hours = ☐ days.

14 70c + 5c + 25c

Mathematical Reasoning

How old are Rosie and David if:

15 Rosie is a teenager whose age is a multiple of 7?	
16 David is also a teenager whose age is a multiple of 5?	

Measurement Time in minutes

Draw the times on these clock faces.

1

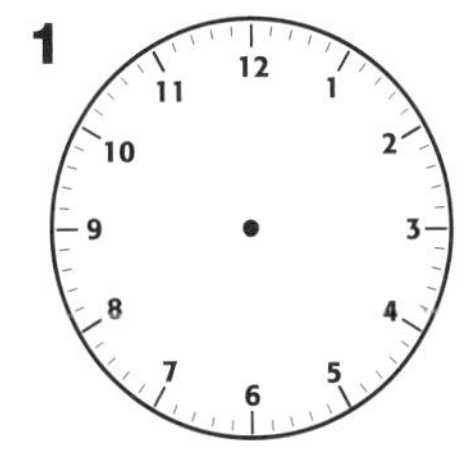

5 to 12

2

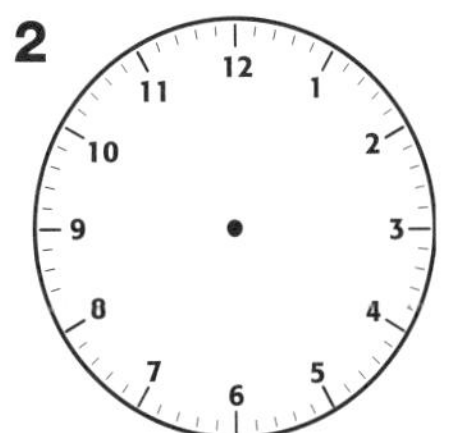

25 past 12

3

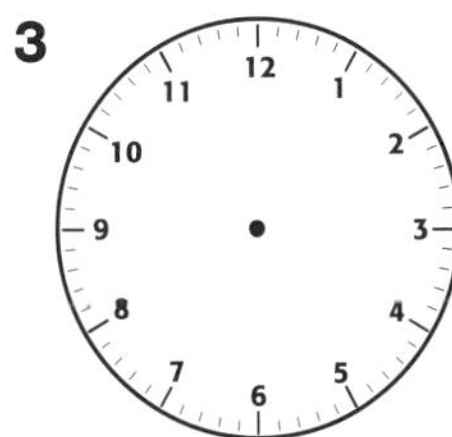

7 o'clock

4

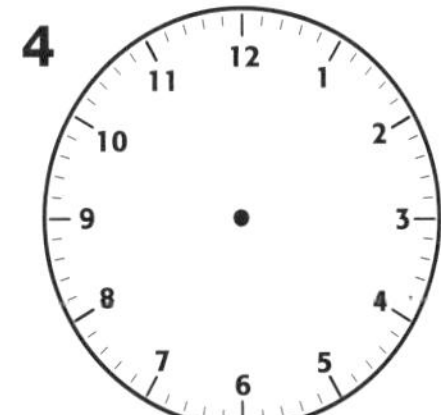

A quarter to 8

5

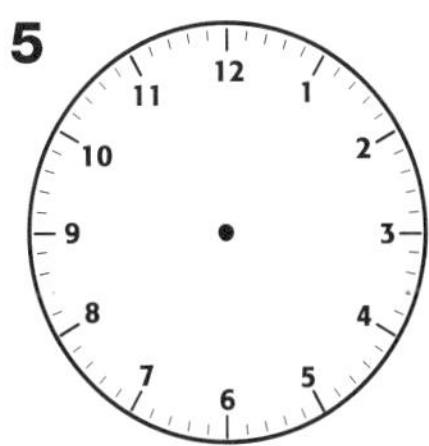

Half past 10

UNIT
16

Number and Algebra

SET 1 Basic

1 17 + 6

2 19 − 3

3 5 × 10

4 16 + 4

5 26 − 4

6 4 × 5

7 17 + 3

8 17 − 3

9 5 × 4

10 Product of 5 and 3

11 Sum of 5 and 10

12 Product of 4 and 5

13 Difference of 8 and 4

14 7 less than 9

15

Luke had $28 but spent $8 on a toy. How much money has he left?

$ ☐

SET 2 Addition with magic squares

The sum of each vertical, horizontal and diagonal line of numbers is the same.

Example

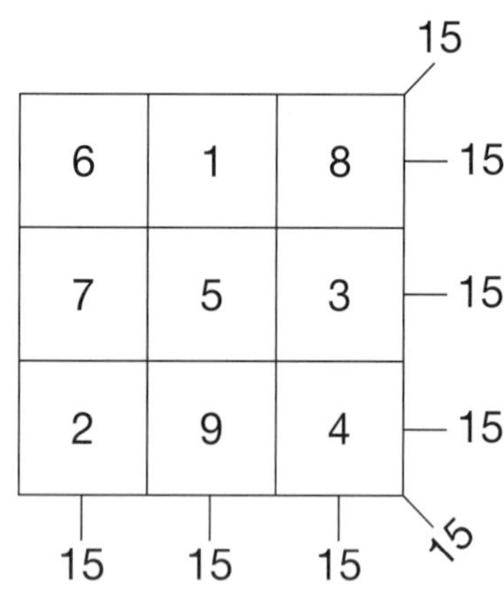

Complete these magic squares.

1

9	4	5
		10
7		3

2

7	2	9
	6	4
	10	

3

	4	11
10	8	6
	12	

4

14	9	10
		15
12		8

Statistics Collecting data

Our class did a survey on the type of pets we had at home. Colour the boxes in the graph to record the pets tallied in our survey.

Pet survey

dog	𝍸 ||||
cat	𝍸 |
bird	||||
fish	𝍸 ||
mice	||

Pets in our class

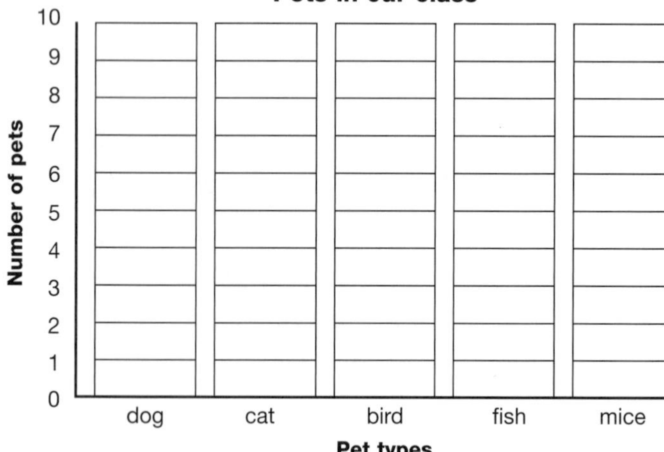

Number and Algebra

SET 3 Multiplication problems

1 3 × 2

2 5 × 2

3 7 × 2

4 9 × 2

5 6 × 2

6 3 × 10

7 5 × 10

8 7 × 10

9 9 × 10

10 6 × 10

11 3 × 5

12 5 × 5

13 7 × 5

14 9 × 5

15 6 × 5

16 8 × 5

17 How many different arrays can you make with 24 counters? ___________
Draw one of your arrays.

Mathematical Reasoning

SET 4 Extension

1 Round 1328 to the nearest 100.

2 How many sides in 14 rectangles?

3 95 = ☐ tens + ☐ ones

4 162 = ☐ hund + ☐ tens + ☐ ones

5 Total of 27 and 600

6 Share 60 among 6.

7 13 more than 122

8 Write 39 in words. ☐

9 What do I add to 35 to make 50?

10 150, ☐, 250, ☐, 350

11 How much are 5 envelopes at 50c each?

12 What odd number comes before 240?

13 Write 7 as an ordinal number.

14 How many hundreds in 375?

15 Jenny bought 8 lollies for 10 cents each. How much change did she get from $2? 

Measurement The kilogram

Colour the correct sets of scales.

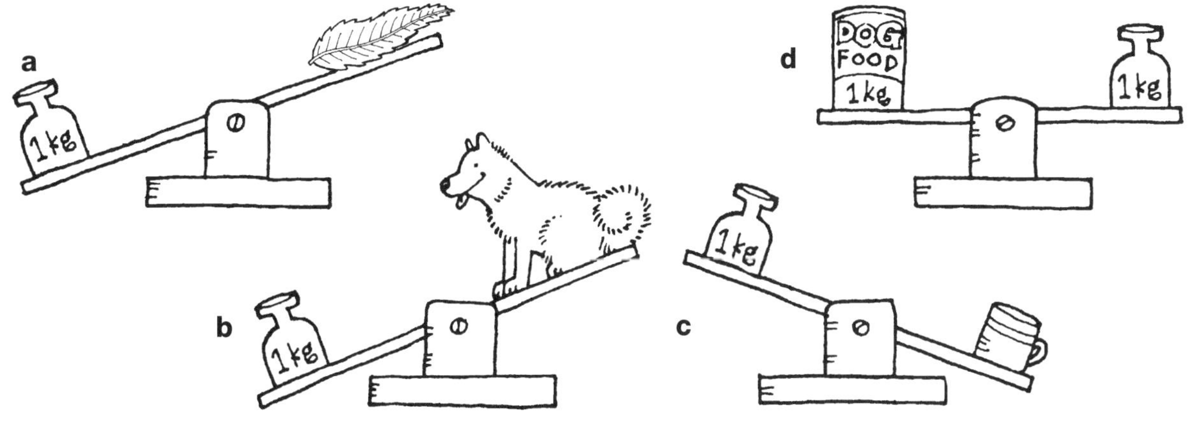

UNIT 17

Number and Algebra

SET 1 Basic

1 6 + ☐ = 10

2 4 times 4

3 Double 6.

4 14 – 10

5 6 × 10

6 5 × 10

7 17 – ☐ = 14

8 17 – ☐ = 7

9 15 + 4

10 What day follows Thursday?

11 10 + 16

12 10, 20, ☐, ☐, 50

13

Complete the grid to show how much Jenny will save in 5 weeks.

Weeks	1	2	3	4	5
Amount	$3	$6	$9		

SET 2 Friendlier calculations

Add these numbers by first adding to the nearest ten.

	Numbers	Becomes	Answer
1	9 + 8	9 + 1 + 7	
2	18 + 7	18 + 2 + 5	
3	48 + 9		
4	64 + 8		
5	36 + 5		
6	45 + 7		

Add these numbers by first adding to the next hundred.

	Numbers	Becomes	Answer
7	98 + 32	98 + 2 + 30	
8	96 + 44		
9	95 + 35		
10	86 + 34		
11	185 + 45		
12	176 + 54		

Number and Algebra Think board/multiplication and addition

Choose between addition or multiplication.

1

Problem

How much would it cost Jarrah to buy 5 soccer balls at $14 each.

Model

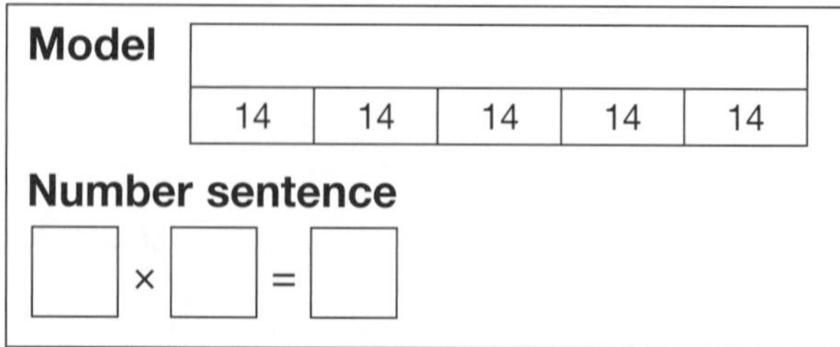

14	14	14	14	14

Number sentence

☐ × ☐ = ☐

2

Problem

How much would it cost Alinta to buy three staplers at $19 each.

Model

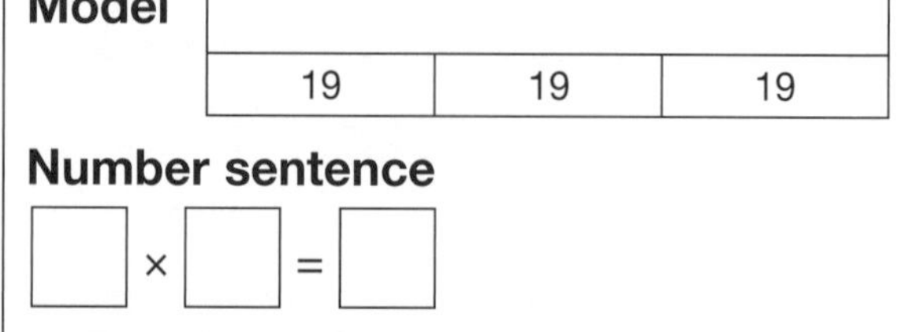

19	19	19

Number sentence

☐ × ☐ = ☐

Number and Algebra

SET 3 Division

1 Seven tickets were bought with $28. How much was each ticket?

2 Twenty-one sausages were cooked for 7 people. How many did each person receive?

3 Complete the grid.

÷	24	32	16	12	8	40	4
4	6			3			

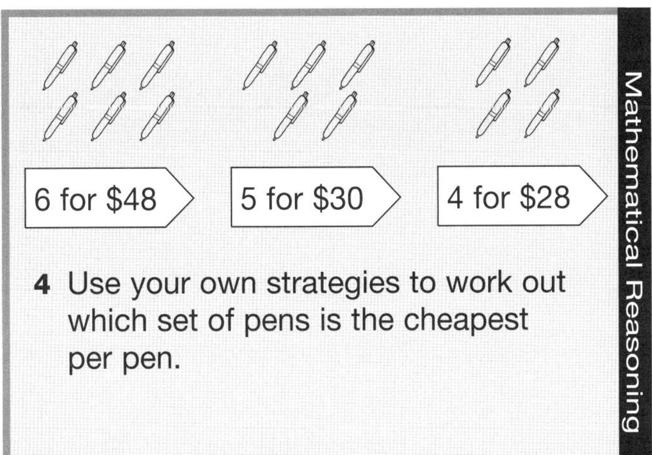

4 Use your own strategies to work out which set of pens is the cheapest per pen.

Mathematical Reasoning

SET 4 Extension

1 What is the sum of 3, 6 and 8?

2 1 litre = ☐ millilitres

3 Which is greater, 2 dozen or 21?

4 Write the greatest number you can using 4, 7, 5.

5 Round 2326 to the nearest 100.

6 How many minutes are in half an hour?

7 4 times zero

8 Estimate an answer to 136 + 141.

9 How many tens are in 894?

10 What is the product of 6 and 2?

11 Share 30 among 6.

12 137, 143, 149, ☐, ☐

13 How much are 2 tickets at 75c each?

14 Twenty centimetres plus 130 cm

15 Circle the date 21 days after 17 May.

MAY						
Su	M	Tu	W	Th	F	S
	1	2	3	4	5	6
7	8	9	10	11	12	13
14	15	16	17	18	19	20
21	22	23	24	25	26	27
28	29	30	31			

JUNE						
Su	M	Tu	W	Th	F	S
				1	2	3
4	5	6	7	8	9	10
11	12	13	14	15	16	17
18	19	20	21	22	23	24
25	26	27	28	29	30	

Measurement Length and width

1 Measure the length of this shape. ____ cm

2 Measure the width of this shape. ____ cm

3 Draw a shape with a length of 6 cm and a width of 2 cm.

Number and Algebra

SET 1 Basic

1 5 + 3

2 15 + 3

3 25 + 3

4 35 + 3

5 12 – 5

6 16 – 5

7 16 – 6

8 16 – 8

9 16 – 10

10 7 × 4

11 9 × 5

12 8 × 4

13 3 × 4

14 Product of 7 and 5

15

John saved $4 each week for 7 weeks. How much did he save?

$ ☐

SET 2 Related multiplication by 2 and 4

1 2 × 2

2 4 × ☐ = 8

3 3 × 4

4 6 × ☐ = 24

5 2 rows of 10

6 10 rows of 4

7 4 × ☐ = 32

8 How many sides on 4 squares?

9 How many legs on 10 birds?

10 How many pencils in 8 boxes of 4 pencils?

Mathematical Reasoning

11 This array was made to show 6 × 4.
Write 4 other multiplication or division facts you can see inside the array.

6	×	4	=	24
			=	
			=	
			=	
			=	

Number and Algebra Think boards/addition

Use the Think boards to solve the problems.

1 How much would it cost Jedda to buy a glue stick and scissors?

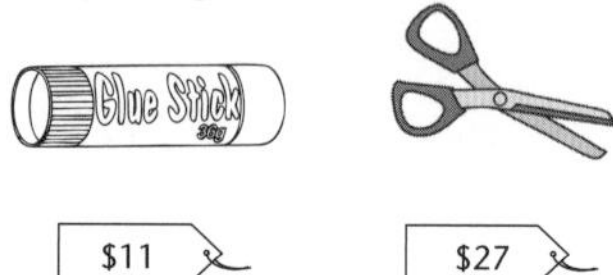

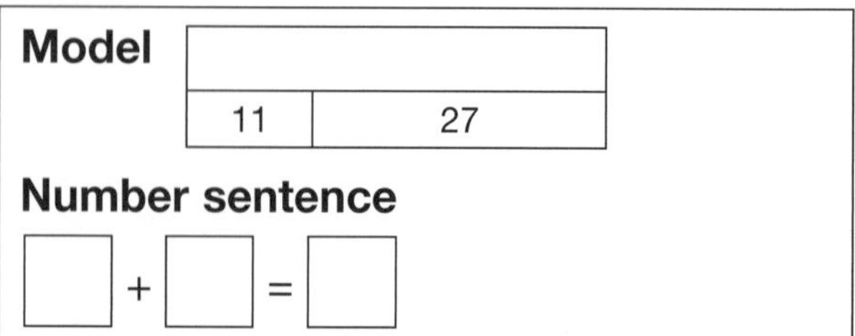

2 How much would it cost Jedda to buy a ruler and a calculator?

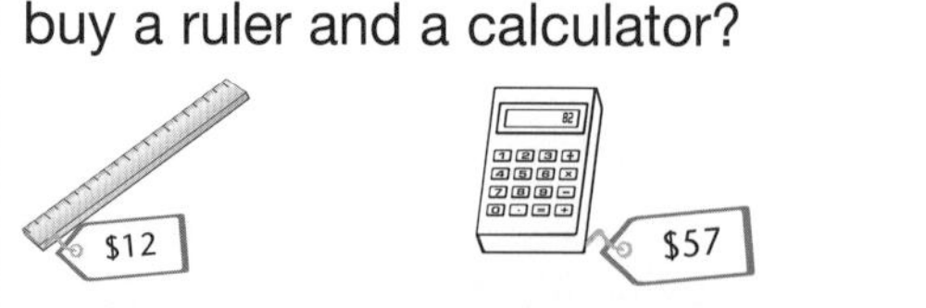

Model

12	57

Number sentence

☐ + ☐ = ☐

Number and Algebra

SET 3 Missing terms in number patterns

Discover the rules and complete the grids.

1

$75	$70	$65			

2

$16	$20	$24			

Find the missing numbers.

3

10	13		19	22		28	31

4

62	58	54		46		38	34

5

0	9	18		36		54	

6

57	65	73		89		105	113

Complete each pattern then write the rule for it.

7

20	30	40	50				

Rule:

8

9	12	15	18				

Rule:

SET 4 Extension

1 How many hundreds are there in 866?

2 $\frac{1}{4}$ of 12

3 How many hours from 9 am to noon?

4 What is the difference between 47 and 23?

5 Peter bought 5 books at $7 each. What was the total cost?

6 4 hundreds + 7 tens + 3 ones

7 Write the largest 3-digit number you can using 7, 9, 1.

8 Write the smallest 3-digit number you can using 7, 9, 1.

9 936 + 4 tens

10 If juice cost $2 a litre, how much would half a litre cost?

Mathematical Reasoning

11 Ms Stevens gives the students in our class 1 sticker each during the week. She gave out 16 stickers on Monday, but only half that number on Tuesday. She continued this pattern for the rest of the week. How many students are in our class?

Space Sides and angles

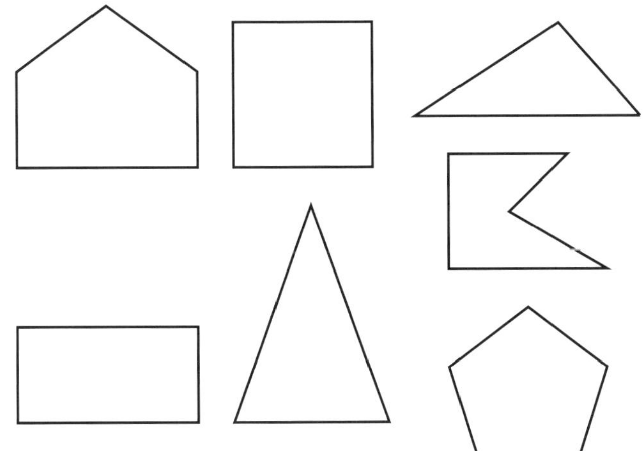

1 Colour the shapes with 3 angles and 3 sides red.

2 Colour the shapes with 4 angles and 4 sides green.

3 Colour the shapes with 5 angles and 5 sides yellow.

UNIT 19

Number and Algebra

SET 1 Basic

1 9 + 4

2 \$10 – \$6

3 Half of 16

4 \$7 + \$9

5 Subtract 6 from 11.

6 4 × 5

7 15 – 3

8 7 less than 18

9 5 × ☐ = 15

10 6 × ☐ = 24

11 Double 12.

12 Sum of 13 and 6

13 21 + 4 + 10

14 Product of 10 and 9

15

SET 2 Numbers to 10000

Show each number on the number expander.

1 6376

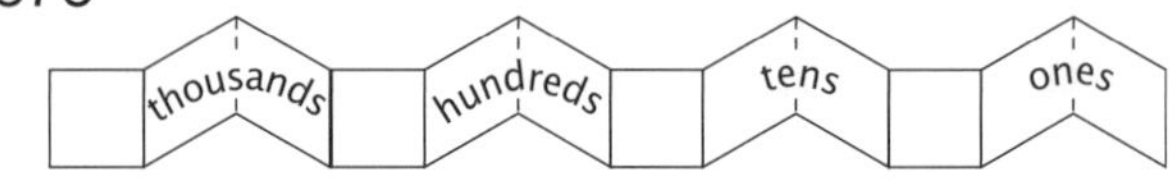

2 5591

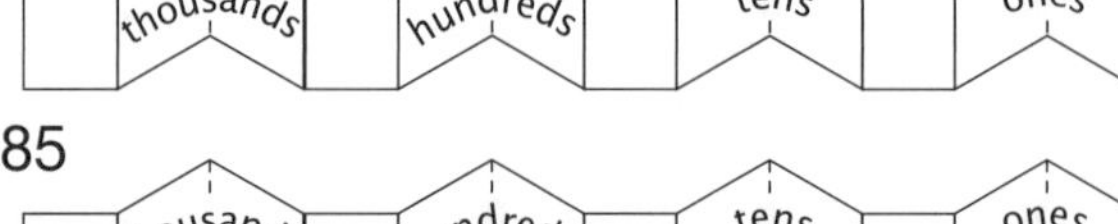

3 8385

thousands | hundreds | tens | ones

4 9296

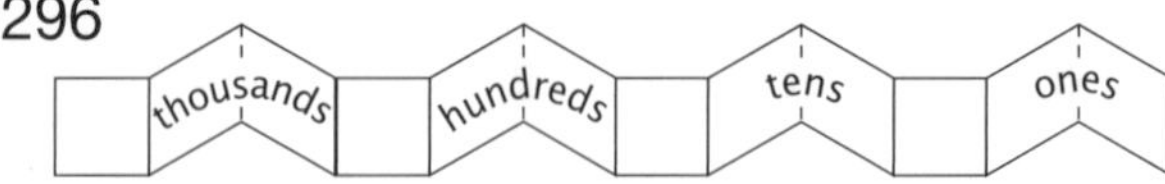

Write these numbers in expanded form.

5 2579 = 2000 + 500 + ☐ + ☐

6 3628 = ☐ + ☐ + ☐ + ☐

7 5196 = ☐ + ☐ + ☐ + ☐

8 7896 = ☐ + ☐ + ☐ + ☐

9 9494 = ☐ + ☐ + ☐ + ☐

10 Order the numbers from least to the greatest.

7437, 8471, 7342 ________________

Space Angles/symmetry

Draw all the lines of symmetry on the shapes.
Name the shapes.

1

____

2

____

3

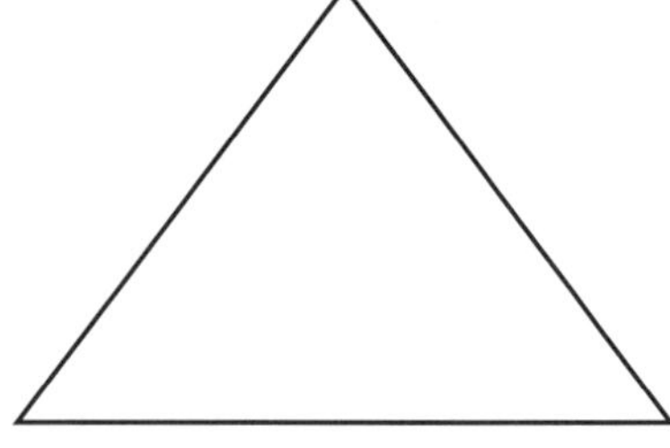

____

Number and Algebra

SET 3 Multiplication facts

1

× 5	
4	
6	
5	
7	
9	
8	

2

× 3	
3	
7	
6	
8	
9	
5	

3

× 2	
1	
2	
4	
9	
6	
5	

Complete the algorithms to identify the winning Bingo card.

4 $3 \times 3 =$

5 $6 \times 3 =$

6 $4 \times 3 =$

7 $5 \times 3 =$

8 $7 \times 3 =$

9 $8 \times 3 =$

10 $6 \times 5 =$

11 $9 \times 5 =$

a

21		15		48	
	30		18		24

b

21		9		15	
	12		18		45

c

21		45		12	
	18		17		48

SET 4 Extension

1 Cost of a $10 cake and a $6 tart

2 How many days in 5 weeks?

3 What is the next odd number after 49?

4 How many minutes from 10 am to 2 pm?

5 Share 60 among 5.

6 $72 + $23

7 6 hundreds plus 7 tens plus 5 ones

8 34 tens plus 9 ones

9 How many 5c coins make 35c?

10 223, 229, 235, ___ , ___ .

Mathematical Reasoning

11 Jenna has forgotten her 3-digit PIN number. The numbers are 4, 5 and 6.

Write all 6 possible combinations on the cards.

My Bank | My Bank | My Bank

My Bank | My Bank | My Bank

Probability Chance

Shade the tag which best describes the chance of the spinner landing on blue.

1

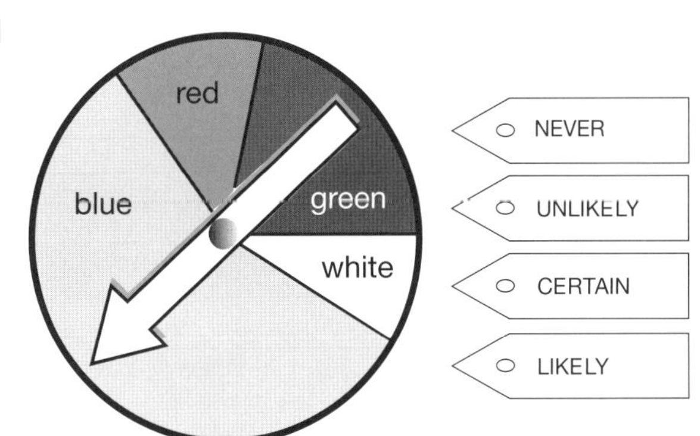

Design a spinner with the colours red, blue and yellow on it. Make sure that red has a higher chance of being landed on than blue.

2

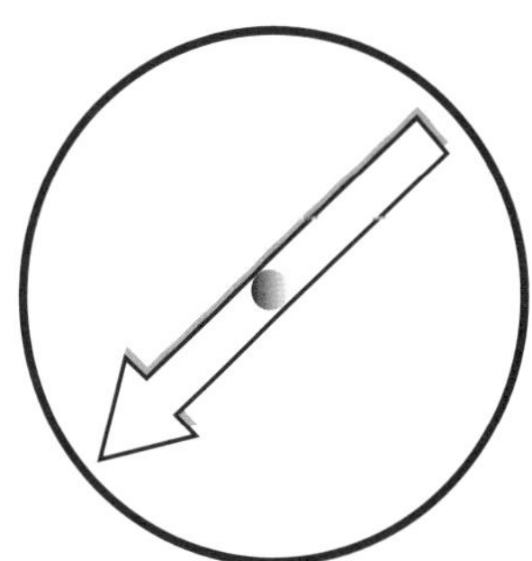

UNIT 20

Number and Algebra

SET 1 Basic

1 23 + 4

2 24 + 5

3 25 + 7

4 27 + 8

5 58 minus 4

6 19 – 5

7 14 – 4

8 18 – 4

9 5×3

10 4×3

11 3×9

12 2×6

13 10×9

14 7 tens and 5 ones

15

SET 2 Division from multiplication

Solve the questions below to find the secret word.

1 $3 \times 8 =$ ☐

2 $24 \div 3 =$ ☐

3 $5 \times 8 =$ ☐

4 $40 \div 5 =$ ☐

5 ☐ $\times 3 = 30$

6 $4 \times 10 =$ ☐

7 $7 \times 3 =$ ☐

8 $21 \div 3 =$ ☐

24	7	2	40	9
I	M	T	E	P

10	8	3	21	12
R	C	S	A	L

9 ___ ___ ___ ___ ___ ___ ___ ___
1 2 3 4 5 6 7 8

10 Joe had 27 marbles, which he needed to sort into groups of 9. How many groups could he make?

11 Jill the tiler needed to put 36 tiles into groups of 3. How many groups did she have?

Number and Algebra Calculating change

Draw strokes under the coins to show four different coin combinations that total $8.95.

		$2	$1	50c	20c	10c	5c
1	$8.95						
2	$8.95						
3	$8.95						
4	$8.95						

Number and Algebra

SET 3 Addition problems

Solve the problems.

1 What is the sum of 27 and 18?

2 What is the total of 53 and 27?

3 Jack had $67. How much more does he need to buy a skateboard that costs $100?

4 Sara saved $57 and $46. What was the total amount she saved?

5 If 76 children attended a concert and 55 of them were seated, how many were standing?

Find the missing digits.

6 37 + ☐ = 50

7 29 + ☐ = 60

8 31 + ☐ = 70

9 30 + 19 + ☐ = 55

Mathematical Reasoning

SET 4 Extension

1 What is the value of 6 in 363?

2 How many hundreds in 573?

3 How many days in 8 school weeks?

4 473c = $ ☐

5 How much more than 130 is 210?

6 What are the factors of 12?

7 Write two thousand, seven hundred and twenty in figures.

8 What is the cost of 3 kg of potatoes at $3 per kg?

9 Jan bought 10 lollies at 17c each. She spent $ ☐

10 30 + 30 + ☐ = 170

11 Write 4 number sentences that equal 75.

☐ + ☐

☐ + ☐

☐ − ☐ = 75

☐ −

Mathematical Reasoning

Space Angles

Draw a line to match each angle on the left to one on the right.

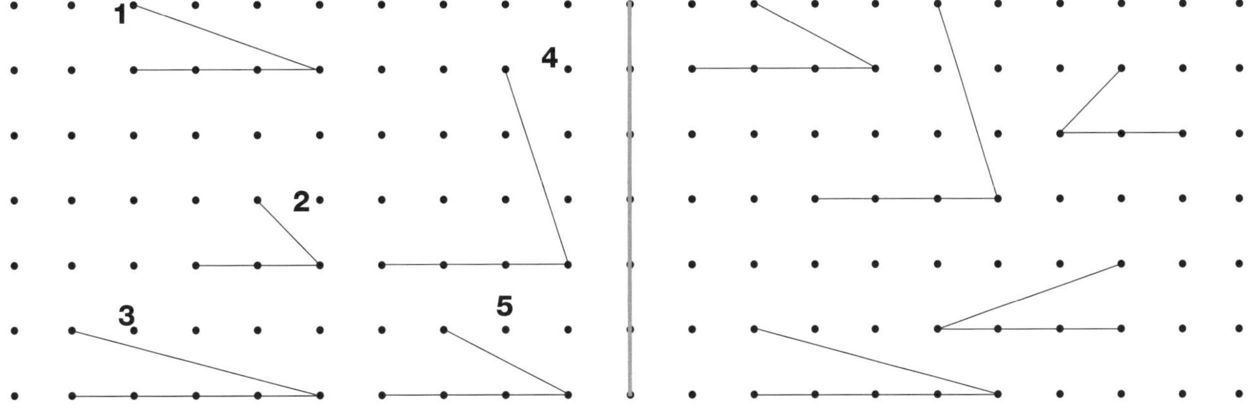

Number and Algebra

SET 1 Basic

1 7 + ☐ = 10

2 Add 7 to 11.

3 Subtract 10 from 50.

4 24 – 4

5 The difference between 47 and 5

6 5 lots of 3

7 7 × 6

8 8 × 4

9 25c + 5c + ☐ = 40c

10 How many hundreds in 224?

11 53 + 6

12 Product of 8 and 5

13 Sum of 8 and 5

14 Days in August

15

How many pairs of socks could be made from 22 socks?

☐ pairs

SET 2 Subtraction–no trading

1

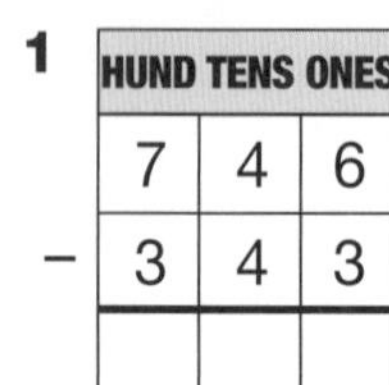

	HUND	TENS	ONES
	7	4	6
–	3	4	3

2

	HUND	TENS	ONES
	6	4	7
–	3	3	6

3

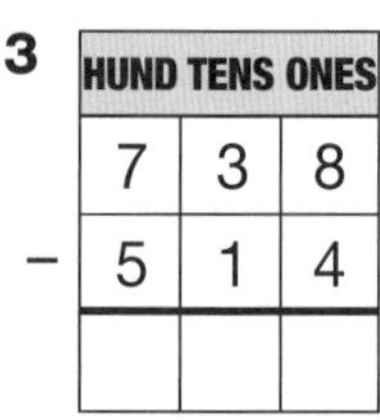

	HUND	TENS	ONES
	7	3	8
–	5	1	4

4

	HUND	TENS	ONES
	4	5	9
–	2	2	8

5

	HUND	TENS	ONES
	5	6	7
–	3	2	6

6

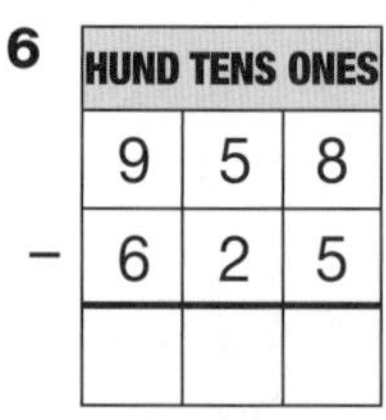

	HUND	TENS	ONES
	9	5	8
–	6	2	5

7

	HUND	TENS	ONES
	7	8	6
–	1	6	6

8

	HUND	TENS	ONES
	8	9	7
–	1	7	7

9

	HUND	TENS	ONES
	8	8	9
–	3	4	4

Mathematical Reasoning

10 Who spent more money when they went shopping?
Liz had $90 and came home with $35.

Mark had $70 and came home with $34.

More money was spent by ____________

11 What was the total amount spent by the children? ____________

Space Describing prisms

Draw a line to match the prisms to their sets of faces.

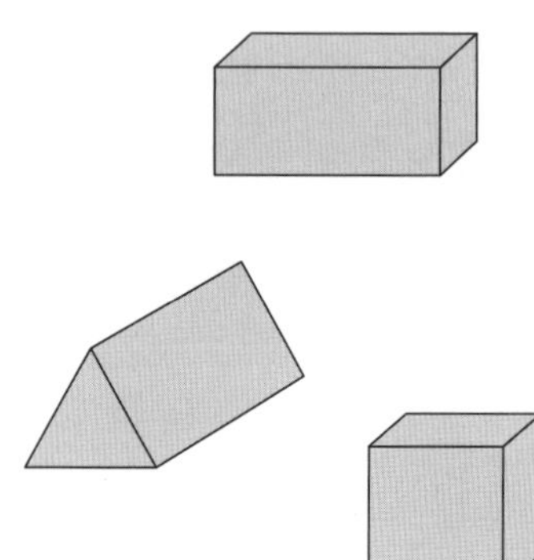

Number and Algebra

SET 3 Rounding to estimate purchases

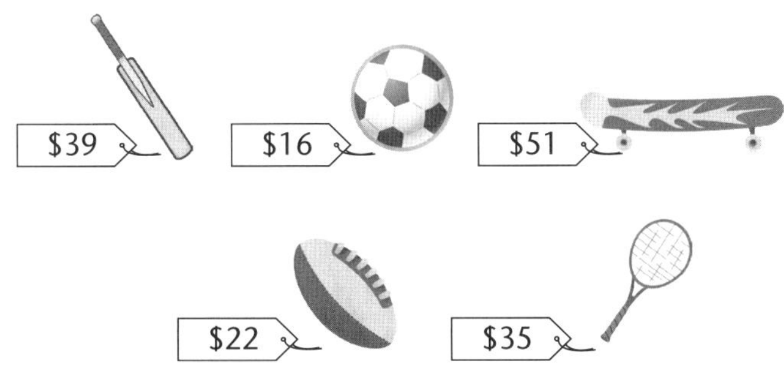

Round off each to the nearest $10 to estimate an answer for the following purchases.

1. a rugby ball and a cricket bat
2. a skateboard and a cricket bat
3. a rugby ball and a tennis racquet
4. a tennis racquet and a skateboard
5. a skateboard and a rugby ball
6. a cricket bat and a tennis racquet
7. a soccer ball and a rugby ball
8. a cricket bat and a soccer ball

SET 4 Extension

1. ☐ × 15 = 30
2. Tom spent 5c and 15c. If he had 20c, how much has he left?
3. How many halves in 3 wholes?
4. 78, 81, 84, ☐ , ☐
5. 3 cakes cost $12. How much did each cake cost?
6. 3 boys shared $60. How much did each boy get?
7. Is a cube a 3D shape?
8. How many 10c coins in $2.30?
9. What is the value of 7 in 7323?
10. Round 2354 to the nearest 1000.
11. How many centimetres in 3 m?
12. If eggs cost $1.20 a dozen, how much for $\frac{1}{2}$ dozen?
13. What is $\frac{1}{4}$ of 48 lollies?
14. Write even numbers between 231 and 239.

Measurement Volume using blocks

Put a cross on all the objects made from 12 cubes.

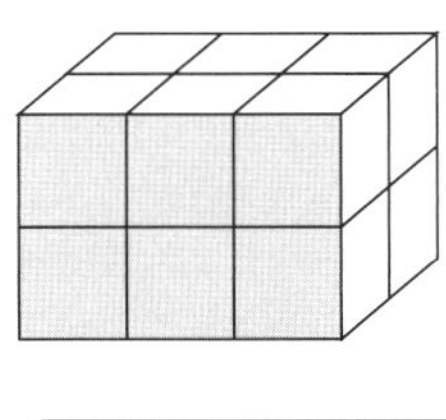
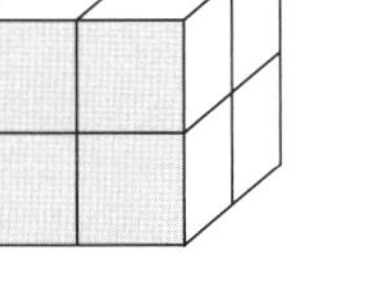
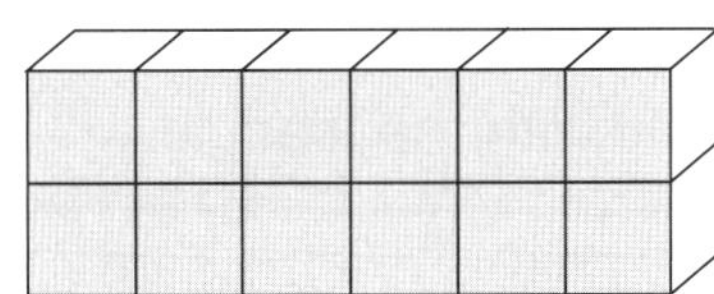
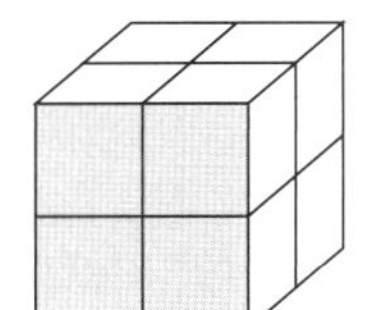
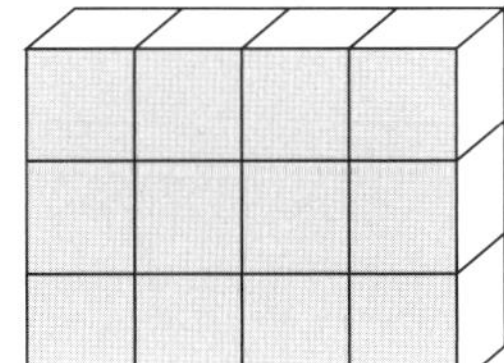
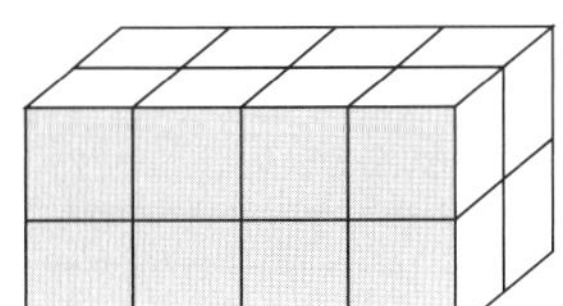
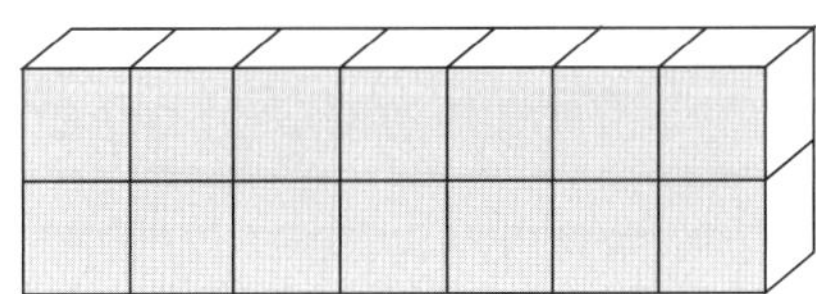

Number and Algebra

SET 1 Basic

1 11 – 2

2 28 – 4

3 36 – 5

4 47 – 6

5 9 + 8 + 7

6 3 + 4 + 3

7 26 take away 7

8 7 + 3 + 8

9 4 × 6

10 3 × 6

11 5 × 6

12 10 × 6

13 18 plus 6

14 Sum of 17 and 12

15

Petra had 6 red lollies, 9 green ones and 6 blue ones. How many did she have altogether?

☐ lollies

SET 2 Trading in addition to 999

1

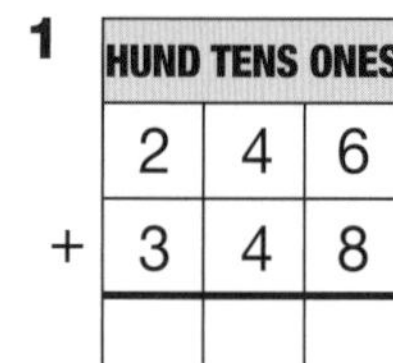

	HUND	TENS	ONES
	2	4	6
+	3	4	8

2

	HUND	TENS	ONES
	3	4	7
+	3	3	6

3

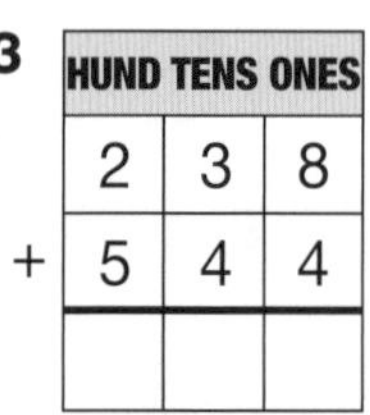

	HUND	TENS	ONES
	2	3	8
+	5	4	4

4

	HUND	TENS	ONES
	4	5	7
+	4	2	8

5

	HUND	TENS	ONES
	5	6	7
+	3	2	6

6

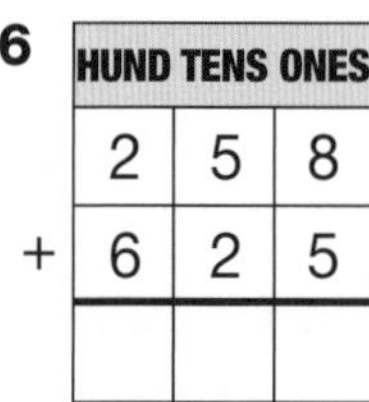

	HUND	TENS	ONES
	2	5	8
+	6	2	5

7

	HUND	TENS	ONES
	7	0	6
+	1	6	6

8

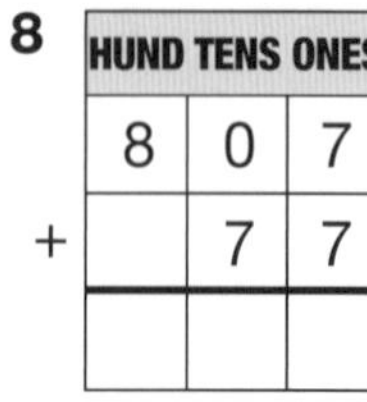

	HUND	TENS	ONES
	8	0	7
+		7	7

9

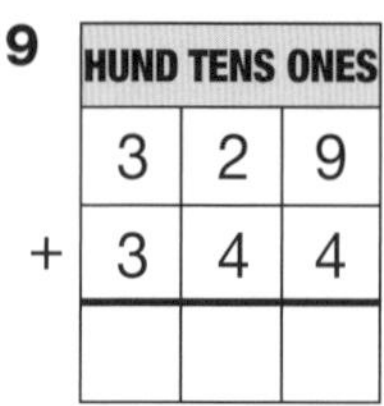

	HUND	TENS	ONES
	3	2	9
+	3	4	4

10 347 boys and 128 girls were members of a soccer club. How many children were members altogether?

11 Jack had $444 and Jenny had $438. How much did they have altogether?

Space Triangles/angles

Draw a line to match the descriptions with the triangles.

I am an equilateral triangle so all my sides are the same length and all of my angles are the same size.

I am an isosceles triangle so 2 of my sides are the same length and 2 of my angles are the same size.

I am a scalene triangle. None of my sides are the same length and none of my angles are the same size.

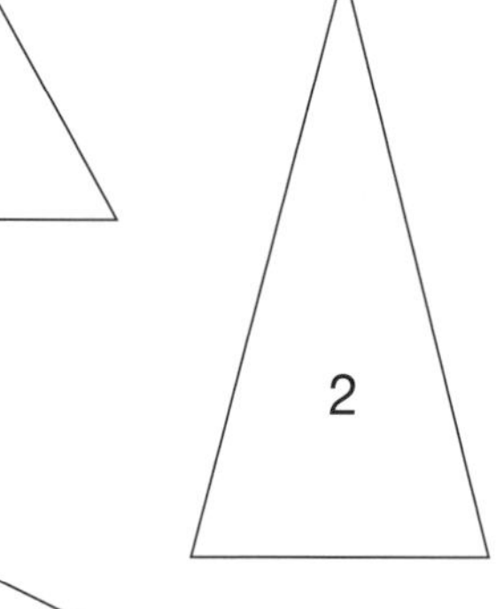

Number and Algebra

SET 3 Division facts and strategies

12 ÷ 3 = ?
Think! 4 × 3 = 12
so 12 ÷ 3 = 4

Use multiplication facts to answer the divisions.

1	10 ÷ 2 = ☐	**7**	28 ÷ 4 = ☐
2	16 ÷ 2 = ☐	**8**	32 ÷ 4 = ☐
3	18 ÷ 2 = ☐	**9**	25 ÷ 5 = ☐
4	9 ÷ 3 = ☐	**10**	35 ÷ 5 = ☐
5	18 ÷ 3 = ☐	**11**	18 ÷ 6 = ☐
6	27 ÷ 3 = ☐	**12**	24 ÷ 6 = ☐

13 12 passengers were shared equally between 4 cars. How many in each car?

14 Two dozen eggs were shared between 4 families. How many did each family get?

15 25 sheep are shared between 5 paddocks. How many sheep are in each paddock?

SET 4 Extension

1 What is the shortest month of the year?

2 How many days in December and September?

3 34, 37, 40, ☐, ☐

4 How many 10c coins in $9.30?

5 If the big hand is on the 12 and the small hand is on the 10, what time is it?

6 How many centimetres in $2\frac{1}{2}$ metres?

7 How many months in 2 years?

8 $5.70 take away 40c

9 185c = $ ☐

10 Share 48 among 6.

11 8 tens minus 8 ones

12 How many sides do 6 octagons have?

13 What is the time 5 hours after 6 pm?

14 50 cm is what fraction of 1 m?

15 2 kilograms = ☐ grams.

16 What is the total number of sides for this group of shapes?

Measurement Digital time

Write a digital time for the times shown on the clocks.

1

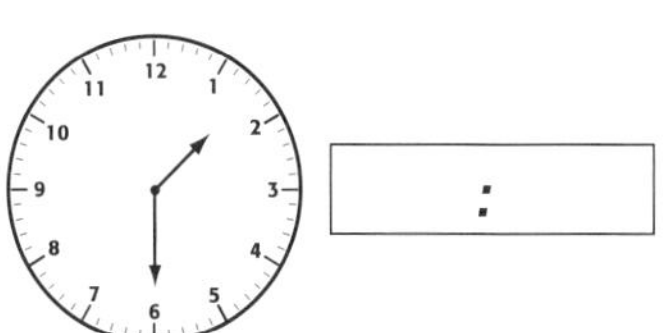

2

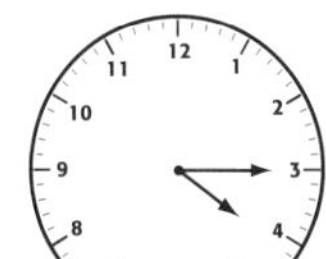

3

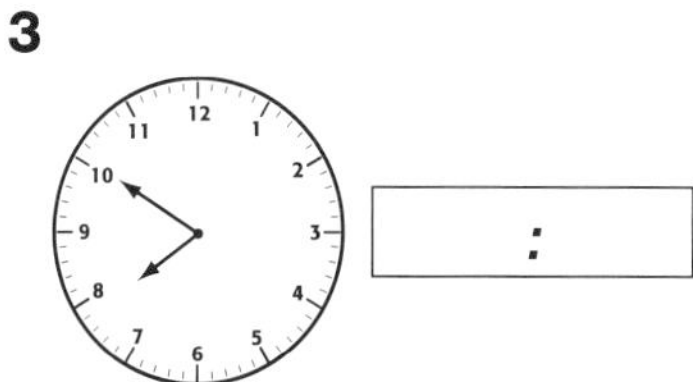

4

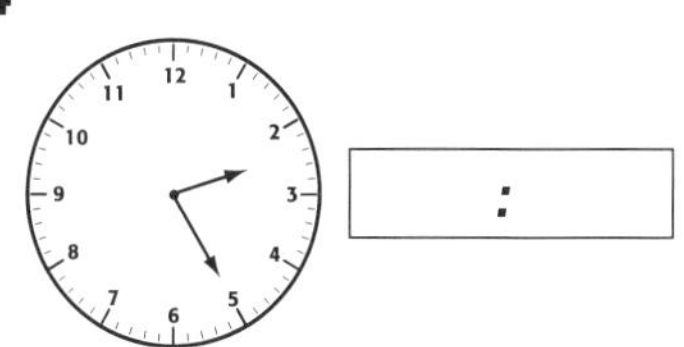

5

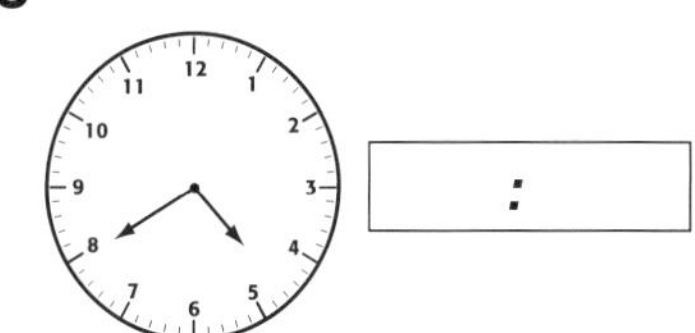

6

UNIT 23

Number and Algebra

SET 1 Basic

1 4 + 8

2 How many cents in $1?

3 $2 + $2 + $2

4 5c + ☐ = 10c

5 20c + 40c

6 15c – 10c

7 11 + 2 + 3

8 $5 × 0

9 $15 + $2

10 60c – 50c

11 Half of 20c

12 3 × 9

13 $1.20 + 5c

14 5 × 9

15 

How many $5 notes are needed to make $40? ☐ notes

SET 2 Trading in subtraction

Complete these subtractions, then find the letter that substitutes for each answer to work out the secret word.

1

	HUND	TENS	ONES
	9	9	1
–	5	6	4

2

	HUND	TENS	ONES
	8	7	2
–	3	4	5

3

	HUND	TENS	ONES
	5	8	3
–	2	5	4

4

	HUND	TENS	ONES
	7	6	2
–	4	3	6

5

	HUND	TENS	ONES
	8	7	3
–	5	5	4

6

	HUND	TENS	ONES
	9	6	1
–	6	3	3

7

	HUND	TENS	ONES
	6	7	0
–	4	4	2

8

	HUND	TENS	ONES
	5	6	3
–	3	4	7

O	T	P	R	C	E	U	M
527	328	326	216	427	228	319	329

9 Secret word

__ __ __ __ __ __ __ __
1 2 3 4 5 6 7 8

Space Angles

Draw examples of these angles.

1 Right angle	2 An angle less than a right angle.	3 An angle greater than a right angle.

Number and Algebra

SET 3 Notes and coins

Draw strokes under the notes to show 5 different note combinations to make up $495.

	$100	$50	$20	$10	$5
1					
2					
3					
4					
5					

6 How much money is in the purse below? $

SET 4 Extension

1 $1.50 + $0.20 = ______

2 $1.45 + $0.40 = ______

3 $1.76 – $0.25 = ______

4 $4.99 – $0.66 = ______

5 Share $2 among 4 people.

6 $3.00 = ☐ c

7 The bus fare is 45c. How much will it cost for 6 trips?

8 How much did I spend if I bought an ice-block for 25c and a cheesestick for 65c?

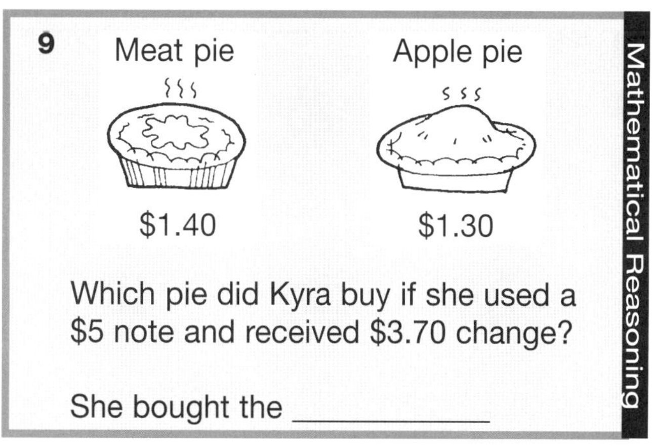

9 Which pie did Kyra buy if she used a $5 note and received $3.70 change?

She bought the ____________

Mathematical Reasoning

Statistics Column graphs

Use the tallied information to complete the graph.

Favourite sports

Soccer	卌 卌 |
Netball	卌 |||
Tennis	|
Softball	||||
Cricket	卌

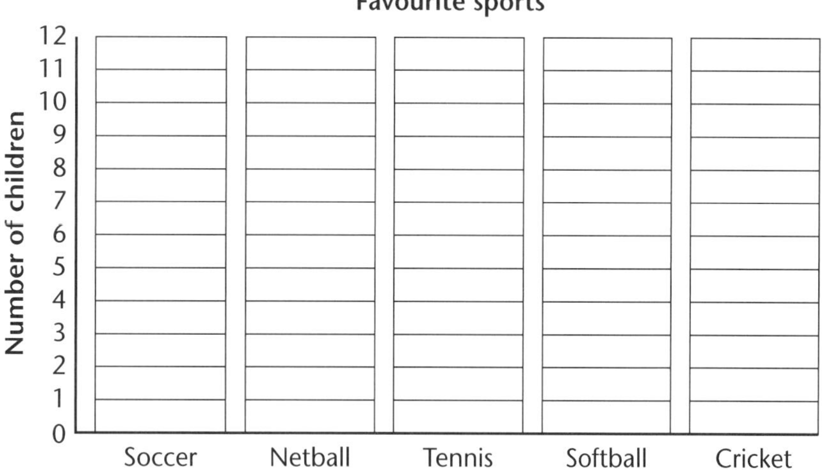

Number and Algebra

SET 1 Basic

1 6 + 8

2 Sum of 6 and 5

3 Add 17 and 6.

4 18 – 5

5 19 – 6

6 20 – 7

7 Half of 50

8 ☐ + 7 = 19

9 10 × 5

10 5 × 6

11 Product of 7 and 5

12 Double 5 plus 10

13 Add 6 to 21.

14 What is the difference between 16 and 8?

15 

Jody planted 5 beds of 6 flowers. How many flowers did she plant?

☐ flowers

SET 2 Making numbers manageable

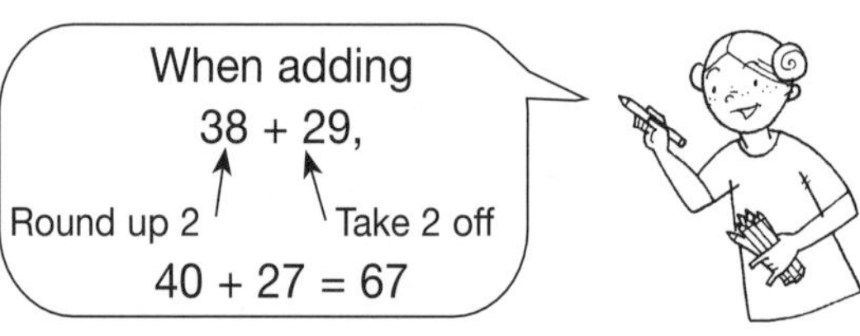

Use the strategy above to find answers to these equations.

	Equation	Make manageable	Answer
1	23 + 38	23 + 40 – 2	
2	32 + 29		
3	18 + 49		
4	25 + 39		
5	27 + 48		
6	46 + 19		
7	48 + 49		
8	53 + 28		

Find answers to the following.

9 39 + 19

10 41 + 38

11 16 + 79

12 24 + 59

13 39 + 69

Statistics Interpreting data

Create a graph from the data for the goals scored.

Name	Goals
Harry	卌 ||
Joe	||||
Sam	卌 卌
Peter	卌 ||||
Sue	|||
Lauren	卌 ||

Goals scored

10, 8, 6, 4, 2, 0

Harry Joe Sam Peter Sue Lauren

Number and Algebra

SET 3 Fifths

What fraction of each shape is shaded?

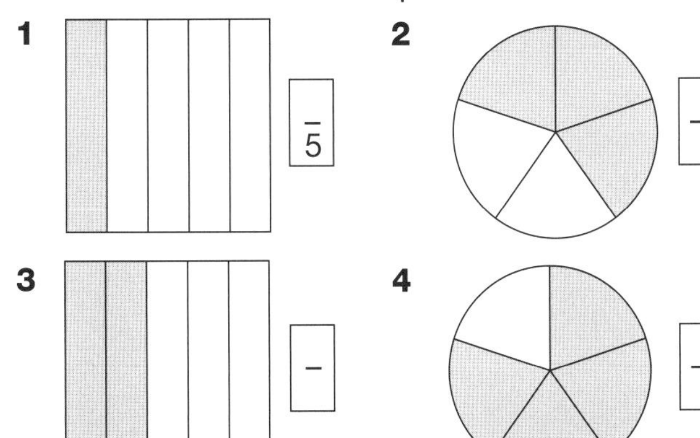

Shade the given fraction.

5

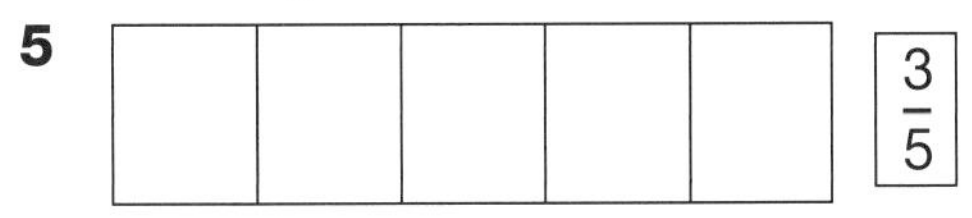

1 whole				
$\frac{1}{2}$	$\frac{2}{2}$			Halves
$\frac{1}{3}$	$\frac{2}{3}$	$\frac{3}{3}$		Thirds
$\frac{1}{4}$	$\frac{2}{4}$	$\frac{3}{4}$	$\frac{4}{4}$	Quarters
$\frac{1}{5}$	$\frac{2}{5}$	$\frac{3}{5}$	$\frac{4}{5}$	$\frac{5}{5}$ Fifths

Use the fraction grid to answer true or false to the statements below.

6 One fifth is larger than one quarter.

7 Three fifths is larger than one quarter.

8 Two thirds is larger than three fifths.

SET 4 Extension

1 How much are 5 books at $9 each?

2 8 hundreds + 7 tens

3 Estimate an answer to 133 + 42.

4 $1\frac{1}{2}$ kg of sugar at $2.00 per kilogram

5 Minutes in $4\frac{1}{2}$ hours

6 Write 272 in words.

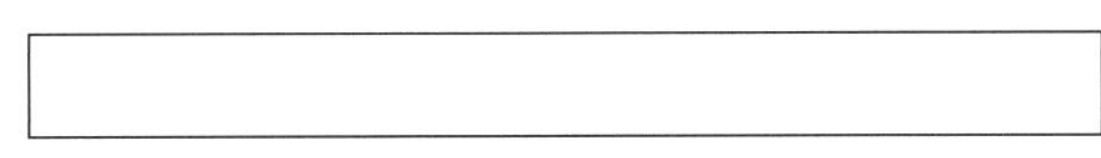

7 How much are 3 combs at 80c each?

8 Value of 3 in 7324

9 What is one third of 90?

10 How many 50c cakes can I buy for $2.50?

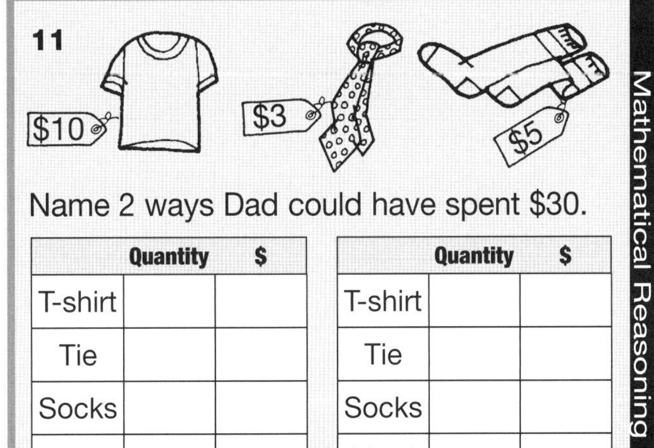

11 Name 2 ways Dad could have spent $30.

	Quantity	$
T-shirt		
Tie		
Socks		
Total		

	Quantity	$
T-shirt		
Tie		
Socks		
Total		

Measurement Time units

What unit of time would you use to measure how long it takes to do the following. (seconds, minutes, hours, months)

1 Building a house __________

2 Brushing your teeth __________

3 Having a shower __________

4 Putting shoes on __________

5 Clothes to dry on the line __________

6 Washing your hands __________

7 Playing a game of football __________

8 Attend school for one day __________

UNIT 25

Number and Algebra

SET 1 Basic

1 5 × 5

2 6 × 0

3 31 + 8

4 ☐ + 5 = 20

5 How many metres in a kilometre?

6 7 times 4

7 18 – 12

8 Half of 20

9 16 + 0 + 4

10 8 tens + 1 more ten

11 4 × 9

12 46 – 6

13 Two quarters of 4

14 60 – 50

15 

Billy saved $3 a week for 6 weeks. How much did he save?

$ ☐

SET 2 Number patterns

Continue the patterns.

1	13	15	17					
2	5	15	25					
3	90	85	80					
4	$\frac{10}{10}$	$\frac{9}{10}$	$\frac{8}{10}$					
5	48	42		30			12	
6	6	12		24				48

Mathematical Reasoning

Create your own patterns.

7								
8								
9								

Probability Chance

Use the spinner to answer the questions below.

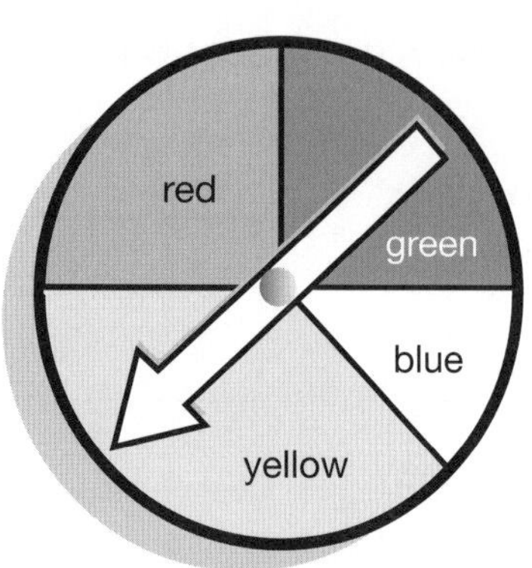

1 Which colour is the spinner most likely to land on?

2 Which colour is the spinner least likely to land on?

3 Which two colours have an equal chance of being landed on?

4 Is the spinner more likely to land on red or green than it is yellow?

Number and Algebra

SET 3 Fractions of a collection

1  $\frac{1}{2}$ of 6 =

2 $\frac{1}{2}$ of 10 =

3 $\frac{1}{2}$ of 12 =

4 $\frac{1}{4}$ of 8 =

5 $\frac{1}{4}$ of 16 =

6 $\frac{1}{3}$ of 6 =

7 $\frac{1}{3}$ of 9 =

8 $\frac{1}{3}$ of 12 =

9 $\frac{1}{3}$ of 15 =

SET 4 Extension

1 How many tens in 360?

2 How many centimetres in $2\frac{1}{2}$ metres?

3 $\frac{1}{4}$ of 60

4 $8 \times \square = 40$

5 Product of 9 and 2

6 How many legs on 6 spiders?

7 Share 12 equally among 3 people.

8 21 + 12 + 9

9 17 tens + 19 ones

10 Share $100 amongst 5.

11 What is the difference between 73 and 8?

12 Amy is half her mother's age. How old is her mother if Amy is 24?

13 How many days are in a leap year?

14 If bracelets are $8.50 each, how much would 2 cost?

15 Which is longer, 1500 m or 2 km?

16 There are 193 pupils at St Mary's Primary School but 48 are on an excursion. How many are still at school?

Measurement Millilitres

Order the capacity of these containers from least to greatest by numbering them 1 to 6.

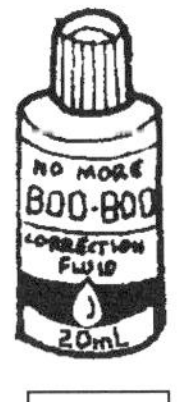

Number and Algebra

SET 1 Basic

1 5×6

2 4×0

3 $21 + 6$

4 ☐ + 3 = 10

5 How many grams in a kilogram?

6 6 times 3

7 19 – 11

8 Half of 12

9 15 + 0 + 2

10 8 tens and 9 ones

11 6×4

12 35 – 4

13 A quarter of 8

14 31 – 21

15 How many days in April and May?

16 Which number sentence has the answer of 21?

❍ 15 + 8 ❍ 30 – 8

❍ 3×6 ❍ 3×7

SET 2 Missing numbers

Supply the missing numbers or symbols.

1 25 + ☐ = 40

2 33 + ☐ = 50

3 7 ☐ 2 = 14

4 29 ☐ 13 = 16

5 7 × ☐ = 35

6 4 ☐ 6 = 24

7 36 ☐ 4 = 9

Mathematical Reasoning

How many ways can you make 48?
One way is shown.

8 88 – 40 = 48

9 ☐ + ☐ = 48

10 ☐ ☐ ☐ = 48

11 ☐ ☐ ☐ = 48

12 ☐ ☐ ☐ = 48

Space Sketching 3D objects

Sketch these objects.

1 Cube	2 Cylinder	3 Square pyramid

Number and Algebra

SET 3 Trading in addition to 999

1

	HUND	TENS	ONES
	3	7	4
+	3	1	7

2

	HUND	TENS	ONES
	5	6	8
+	1	2	4

3

	HUND	TENS	ONES
	6	4	9
+	2	1	6

4

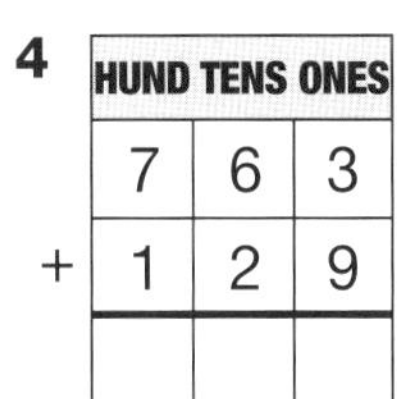

	HUND	TENS	ONES
	7	6	3
+	1	2	9

5

	HUND	TENS	ONES
	5	3	6
+	4	4	5

6

	HUND	TENS	ONES
	3	4	7
+	3	3	7

7

	HUND	TENS	ONES
	4	4	8
+	3	3	4

8

	HUND	TENS	ONES
	5	7	9
+		1	6

9

	HUND	TENS	ONES
	5	2	8
+	2	0	8

10 What is the total cost of a $437 TV set and a $238 carpet?

11 Jack had a bag of 374 marbles and Kim had a bag of 217. How many marbles did they have altogether?

SET 4 Extension

1 How many tens in 450?

2 How many grams in a quarter of a kilogram?

3 $\frac{1}{4}$ of 16

4 $9 \times \square = 45$

5 Product of 6 and 3

6 6 lollies cost 30c, how much for 4?

7 Share 48 among 4.

8 16 + 4 + 160

9 Which is larger: $\frac{1}{4}$ or $\frac{1}{10}$?

10 7 tens + 21 ones

11 What is the difference between 642 and 9?

12 Xia is 9 and Chi is twice that age. How old is Chi?

13 How many days are there in a normal year?

14 If each watch costs $16, how much would 4 watches cost?

15 Round 2431 to the nearest 100.

Measurement Time in minutes

Write the digital time that is 20 minutes later than the time shown on the clock faces.

1

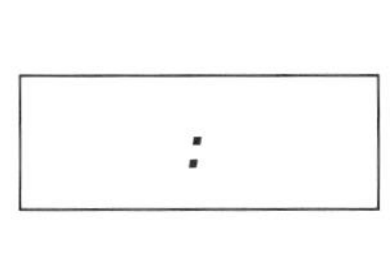

2

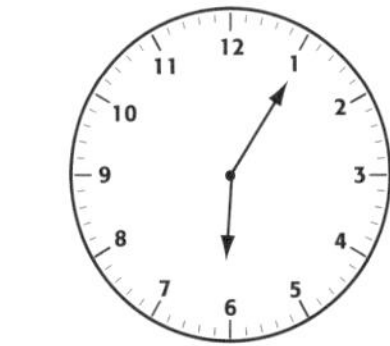

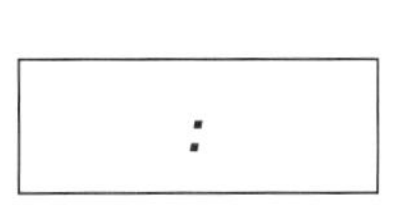

3

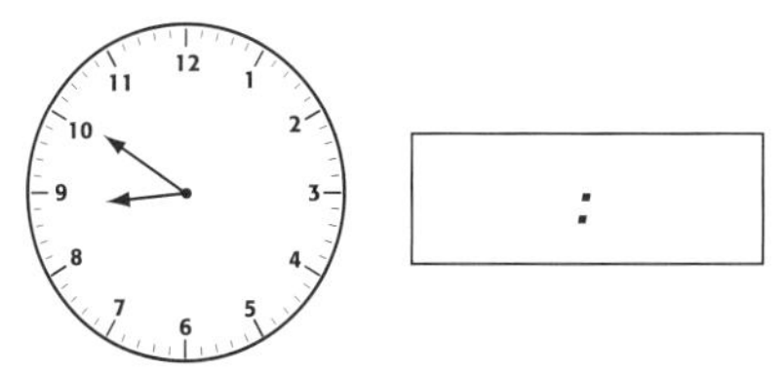

4

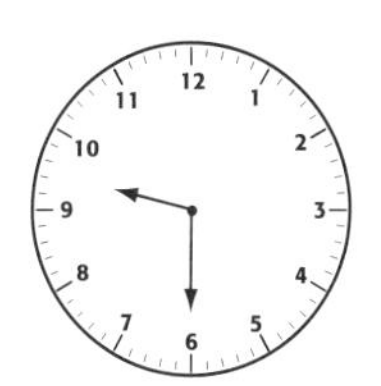

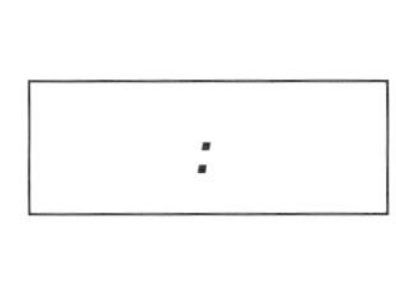

5

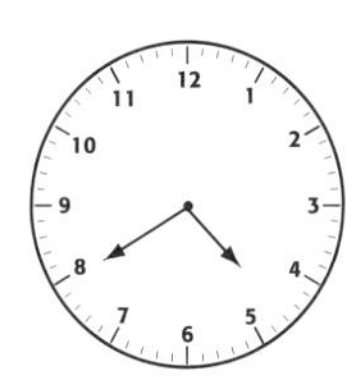

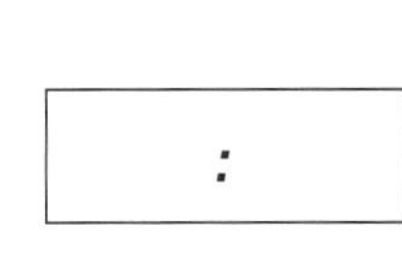

6

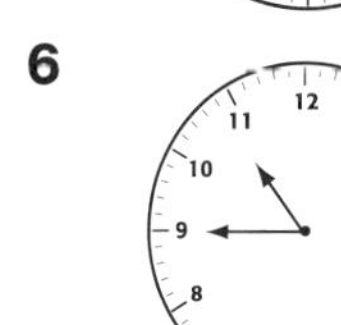

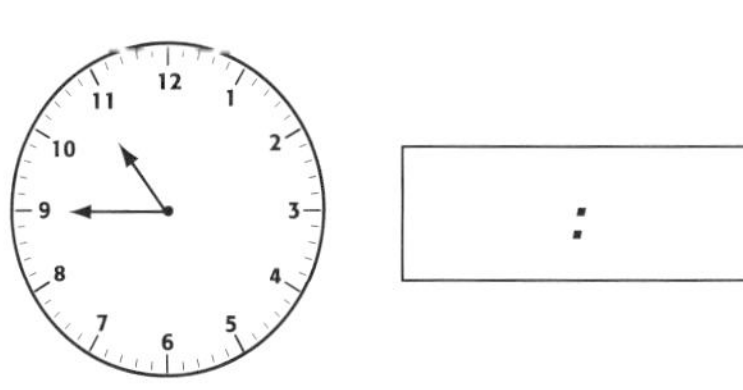

UNIT 27

Number and Algebra

SET 1 Basic

1 40 + 7

2 11 – 8

3 8 + 9 + 2

4 Half of 20c

5 Product of 5 and 8

6 Is 25 an odd number?

7 10 × 2

8 8 × 4

9 150 – 1

10 26 + 13

11 Product of 3 and 4

12 Is December a summer month?

13 21 – 10

14 4 lots of 6

15

Toula planted 3 rows of trees with 6 trees in each row. How many trees did she plant?

☐ trees

SET 2 Multiplication facts

1 7 × 1	8 6 × 5
2 6 × 2	9 6 × 10
3 3 × 3	10 8 × 3
4 5 × 4	11 8 × 4
5 5 × 5	12 8 × 5
6 3 × 6	13 9 × 10
7 7 × 4	14 8 × 0

Mathematical Reasoning

15 Multiply the numbers in the magic square by 5 to create a new magic square. Record the total of the new magic square.

4	9	2
3	5	7
8	1	6

15

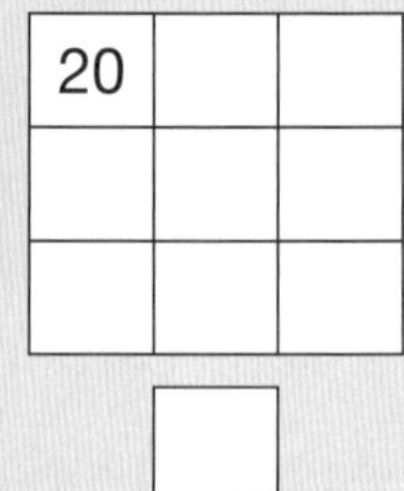

☐

16 Would it still be a magic square if you multiplied by 6?

Space Faces, edges and vertices

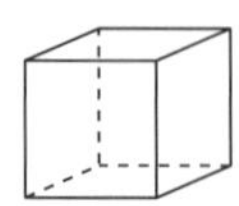

cube

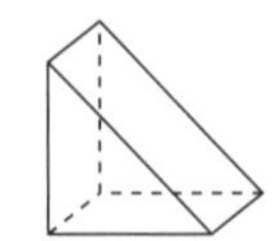

triangular prism

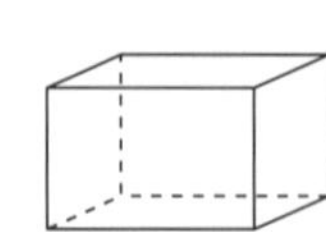

rectangular prism

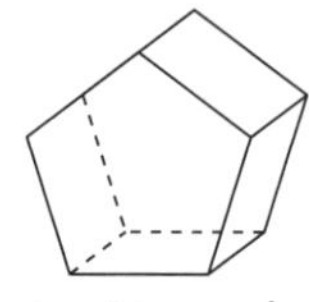

pentagonal prism

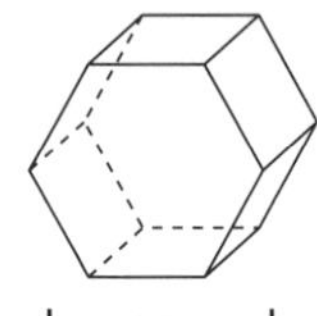

hexagonal prism

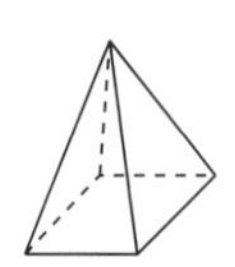

square pyramid

1 How many faces are on a cube?

2 How many faces are on a triangular prism?

3 How many faces are on a rectangular prism?

4 How many faces are on a hexagonal prism?

5 How many edges on a square pyramid?

6 How many edges on a rectangular prism?

7 How many edges on a hexagonal prism?

8 How many vertices on a cube?

9 How many vertices on a square pyramid?

10 How many vertices on a pentagonal prism?

Number and Algebra

SET 3 Extending multiplication facts

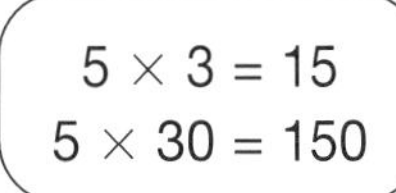

Extend these facts.

1 5×2

2 5×20

3 8×2

4 8×20

5 4×3

6 4×30

7 7×3

8 7×30

9 4×5

10 4×50

11 7×5

12 7×50

13 3×6

14 3×60

15 Bill bought 30 books at $8 each. How much did he spend altogether?

SET 4 Extension

1 \$5.63 + \$0.17 = ______

2 188 km + 14 km = ______

3 46 hours − 26 hours = ______

4 295 cm − 124 cm = ______

5 How much are 3 kg of potatoes at \$2.50 kg?

6 Peter has \$27 for his parents' present and his sister has \$92. How much do they have altogether?

Mathematical Reasoning

7 Place the numbers 6, 2, 1 and 7 into the shapes so that each line adds to 13.

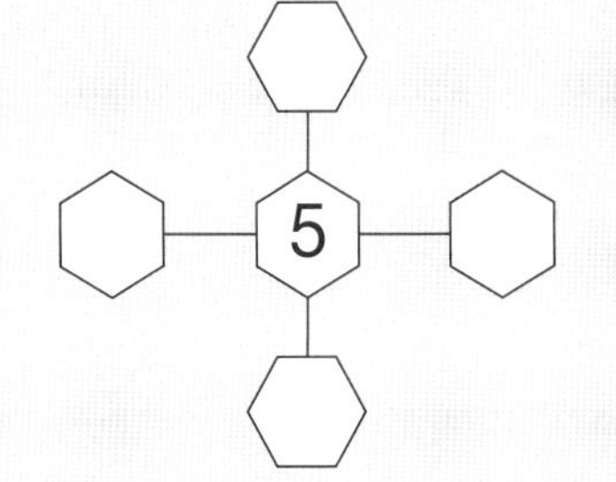

Probability Chance

Colour the marbles in the bag to make the following statements true.

1 A red marble is more likely to be drawn out of the bag.

2 A green marble is least likely to be drawn out of the bag.

3 A blue marble will never be drawn out of the bag.

UNIT
28

Number and Algebra

SET 1 Basic

1 Sum of 6 and 7

2 3 + ☐ = 9

3 13 + ☐ = 20

4 59 + 3

5 48 – 5

6 99 – 9

7 2 × 5

8 4 × 5

9 ☐ × 3 = 18

10 5 × 3

11 7 × 3

12 8 × 3

13 How many fives make 10?

14 How many twos make 20?

15

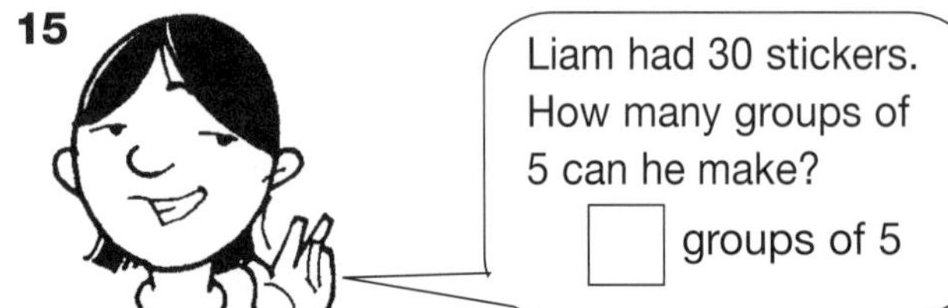

SET 2 Partitioning numbers

Write the number represented by abacuses.

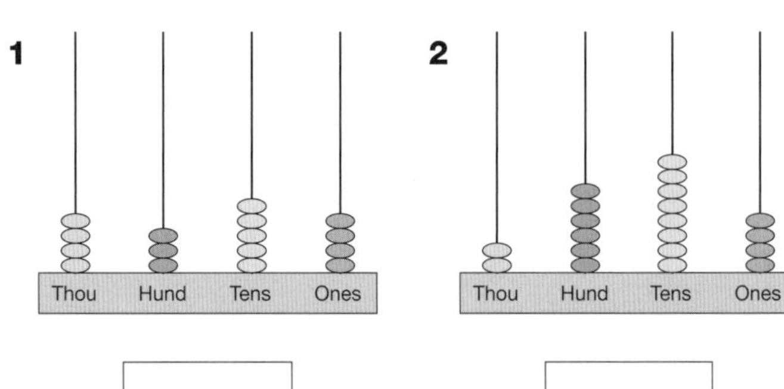

3 Model the number 7612 on the abacus below.

4 Add 6 to the number on the abacus.

5 Record the number you have made in the box below.

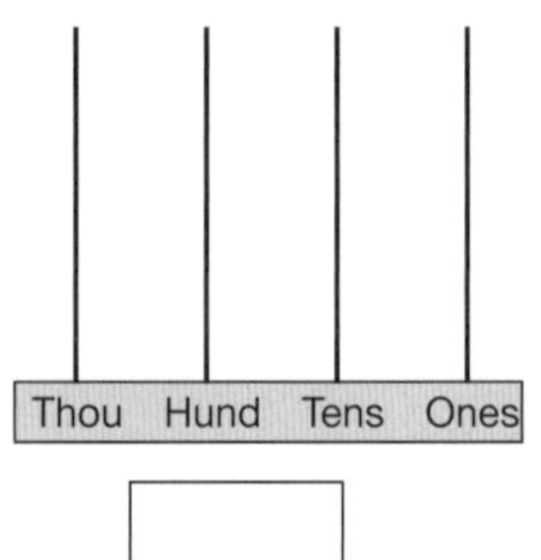

Space Angles and shapes

Join the dots to form the shapes then name them.

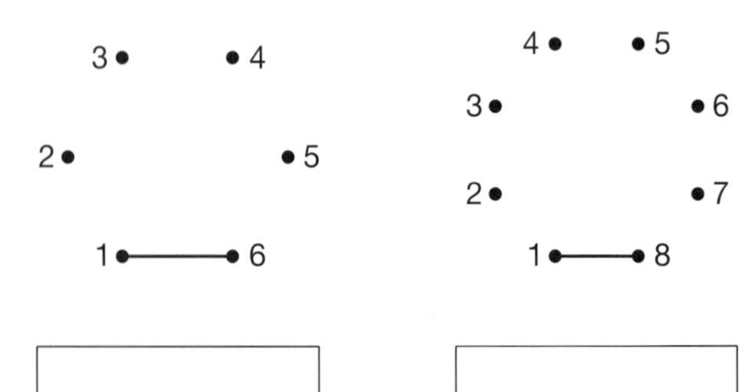

Number and Algebra

SET 3 Commutative property

Write answers for the pairs of additions.

1	8 + 7 =	7 + 8 =
2	9 + 6 =	6 + 9 =
3	12 + 13 =	13 + 12 =
4	12 + 9 =	9 + 12 =
5	20 + 15 =	15 + 20 =
6	15 + 8 =	8 + 15 =
7	25 + 13 =	13 + 25 =
8	40 + 29 =	29 + 40 =

9 Does it matter in what order numbers are added?

Complete each multiplication pair.

10	5 × 2 =	2 × 5 =
11	6 × 2 =	2 × 6 =
12	3 × 5 =	5 × 3 =
13	4 × 5 =	5 × 4 =
14	7 × 5 =	5 × 7 =
15	6 × 4 =	4 × 6 =

16 Does it matter in what order numbers are multiplied?

SET 4 Extension

1 ☐ × 20 = 80

2 Jim spent 7c and 13c. If he had 50c, how much has he left?

3 61, 65, 69, ☐, ☐

4 Three cakes cost $9. How much for 4?

5 Is a prism a 2D shape?

6 How many 50c coins make $12.00?

7 182 – 5

8 Write odd numbers between 82 and 90.

9 How many centimetres in $7\frac{1}{2}$ m?

10 Write the number 9 less than 480.

11 How many angles has a square?

12 286c = $ ☐

13 What are the factors of 16?

14 Double 4 then double again.

15 If eggs cost $1.20 a dozen, how much for $4\frac{1}{2}$ dozen?

16 Share 33 among 5.

Measurement Kilograms

Order these items from lightest to heaviest by numbering them from 1 to 5.

☐

☐

☐

☐

☐

UNIT 29

Number and Algebra

SET 1 Basic

1 9 + ☐ = 16

2 20c – 15c

3 3 × 3

4 16 + 3 + 5

5 Half of 24

6 17 – 6

7 29 + 7

8 100 – 30

9 5 × 10

10 Sum of 21 and 8

11 5c + 5c + 5c

12 7 × 4

13 $5 = ☐ cents.

14 20 minus 13

15

Michael had 6 lots of 10 popsticks. How many popsticks did he have altogether?
☐ popsticks

SET 2 Number patterns

Follow the rules to complete the patterns.

	Rule							
1	+ 5	35	40	45	50			
2	– 5	80	75	70	65			
3	– 5	93	88	83	78			
4	+ 7	10	17	24	31			
5	+ 15	0	15	30	45			
6	– 10	145	135	125	115			
7	– 15	150	135	120	105			
8	+ 30	0	30	60	90			

Measurement am and pm time

Draw the times on the clock faces then write them digitally using am and pm notation.

Ten past seven in the morning	Half past eight in the evening	Twenty past eleven in the morning	Twenty to five in the afternoon
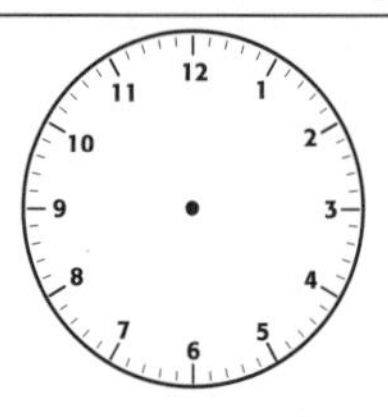	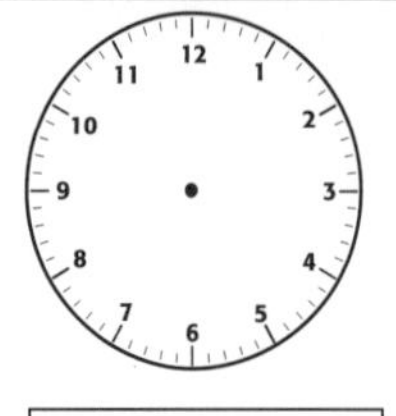	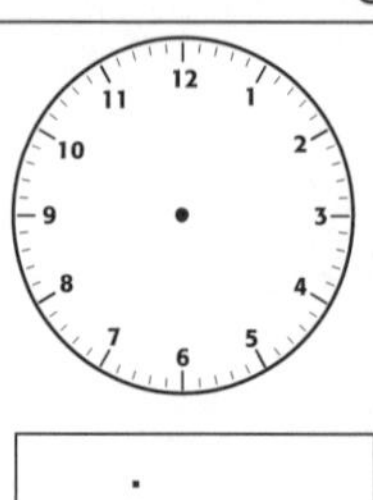	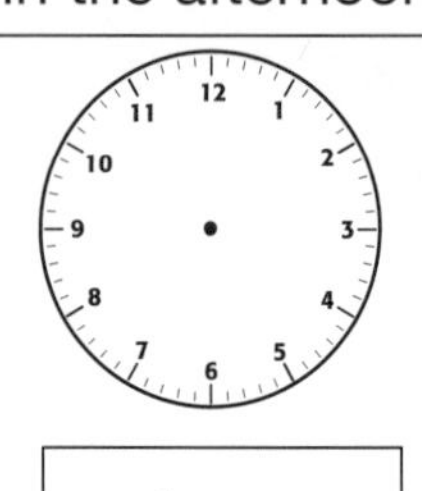
7:10 am	:	:	:

Number and Algebra

SET 3 Fractions of a collection

There are six marbles.

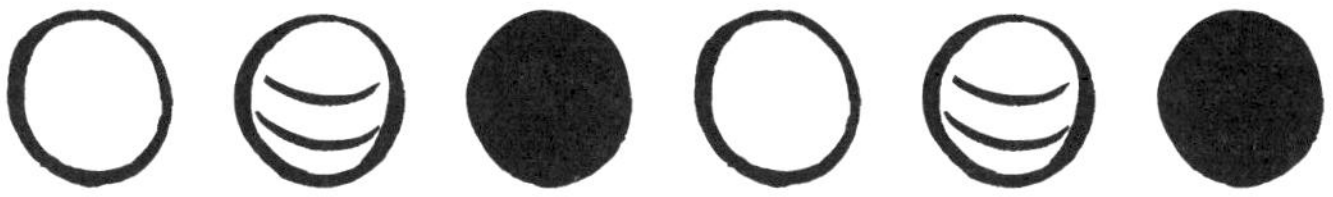

How many marbles are there in:

1 $\frac{2}{3}$ of the collection ______

2 $\frac{1}{2}$ of the collection ______

3 $\frac{1}{3}$ of the collection ______

4 $\frac{2}{6}$ of the collection ______

There are eight keys in a collection.

How many keys are there in:

5 $\frac{1}{2}$ of the collection ______

6 $\frac{1}{4}$ of the collection ______

7 $\frac{3}{4}$ of the collection ______

8 $\frac{5}{8}$ of the collection ______

SET 4 Extension

1
$$\begin{array}{r} 466 \\ 277 \\ +\ \ 43 \\ \hline \\ \hline \end{array}$$

2
$$\begin{array}{r} 324 \\ 28 \\ +\ 639 \\ \hline \\ \hline \end{array}$$

3
$$\begin{array}{r} 424 \\ -318 \\ \hline \\ \hline \end{array}$$

4
$$\begin{array}{r} 645 \\ -\ 328 \\ \hline \\ \hline \end{array}$$

5 Which is smaller, $\frac{1}{2}$ or $\frac{1}{8}$?

6 Which is larger, $\frac{1}{2}$ or $\frac{3}{8}$?

7 325, 350, 375, ☐ , ☐

Mathematical Reasoning

Complete these patterns.

8 (5)—(10), (40)—( )

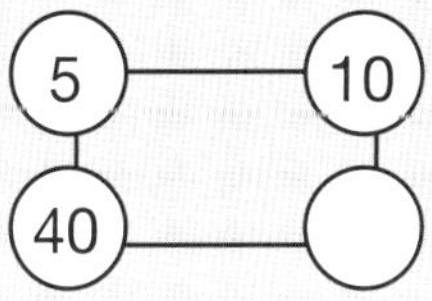

9 (1)—( ), ( )—(16)

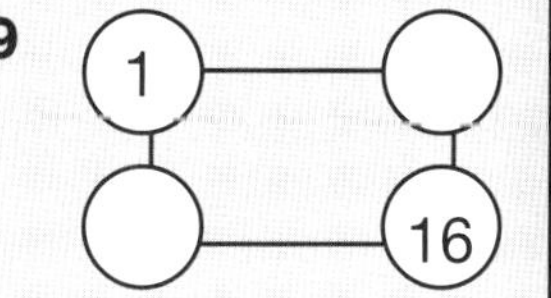

Create a pattern of your own.

10

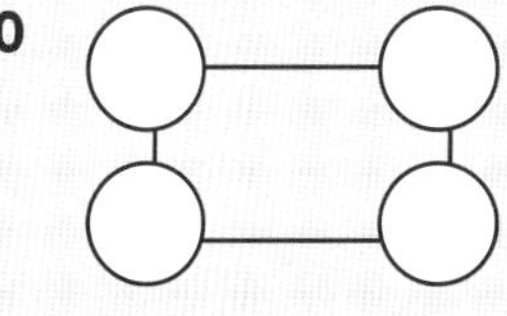

Probability Chance experiments

Here are the results of spinning this spinner 20 times.

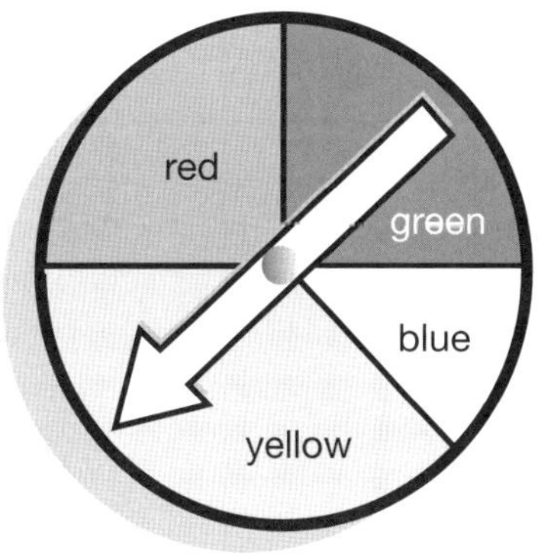

Spinner sections: red, green, blue, yellow

red	~~||||~~
green	||||
blue	||
yellow	~~||||~~ ||||

1 What colour occurred the most often?

______

2 Do you think the spinner is more likely to land on red than on blue? ______

3 a Which colour has the least chance of occurring? ______

b Do the results show this?

______

UNIT 30

Number and Algebra

SET 1 Basic

1 9 + ☐ = 15

2 \$20 – \$7

3 3 + 7 + 3 + 7

4 22 + 8

5 6×3

6 Sum of 24 and 9

7 4×4

8 Six lots of 6

9 27 – 6

10 9×5

11 Four rows of 8

12 8×3

13 Double 9.

14 Product of 4 and 9

15

SET 2 Trading in subtraction to 999

1

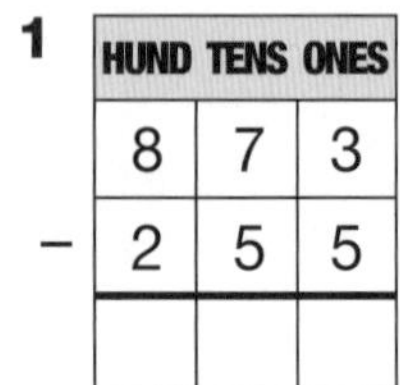

	HUND	TENS	ONES
	8	7	3
–	2	5	5

2

	HUND	TENS	ONES
	9	6	2
–	3	3	7

3

	HUND	TENS	ONES
	8	7	4
–	6	2	8

4

	HUND	TENS	ONES
	8	2	8
–	4	4	3

5

	HUND	TENS	ONES
	9	3	7
–	5	6	2

6

	HUND	TENS	ONES
	7	2	4
–	3	3	1

7

	HUND	TENS	ONES
	8	2	4
–	3	1	6

8

	HUND	TENS	ONES
	9	6	1
–		4	3

9 What is the difference between \$867 and \$345? 

10 How much does Adam need to save if he has \$173 and he wants to buy an \$850 go-cart?

Measurement Benchmarks

Convert these benchmark measurements into metric units.

1 4 medicine glasses _____ mL

2 2 tomato sauces _____ mL

3 5 teaspoons _____ mL

4 6 tablespoons _____ mL

5 $1\frac{1}{2}$ milk cartons _____ mL

6 $2\frac{1}{2}$ cups _____ mL

7 5 tomato sauces _____ mL

8 3 milk cartons _____ mL

medicine glass	tomato sauce	milk carton
	TOMATO SAUCE 500 mL	MILK 1 LITRE
50mL	500mL	1 litre
teaspoon	tablespoon	Cup
5mL	20mL	250mL

Number and Algebra

SET 3 Problems

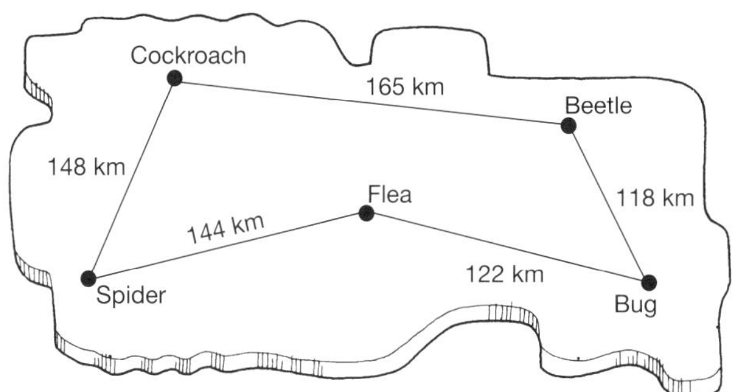

How many kilometres from:

1 Spider to Bug via Flea?
2 Spider to Beetle via Cockroach?
3 Cockroach to Bug via Beetle?
4 Cockroach to Flea via Spider?
5 Cockroach to Flea via Beetle and Bug?

Solve the problems.

6 Ben scored 135 and 146 in two innings. What was his total score for the match? ☐

7 Tom had \$53 and Jill had \$35. How much money did they have altogether? ☐

8 Sally has 2 dozen eggs and Kim has 27 eggs. How many eggs do they have altogether? ☐

SET 4 Extension

1 What is the value of 4 in 743?
2 What is the value of 6 in 637?
3 What digit represents thousands in 3504?
4 How many hours from noon to 6 pm?
5 407 = ☐ hund + ☐ tens + ☐ ones.
6 Write the largest number you can using 1, 3, 2.
7 How many days in spring?
8 What 2 coins make 70c?
9 7 stickers at 15c each.
10 How many halves in $3\frac{1}{2}$?

Mathematical Reasoning

Find pairs of numbers that produce these sums and products. The first one has been done for you.

	Number	Number	Sum	Product
11	3	2	5	6
12			7	12
13			15	50
14			13	36
15			13	40

Probability Chance

Draw a line to match the chance of the dice landing on these faces to a label.

1 The dice lands on a 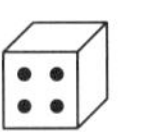

2 The dice lands on an even number. 

3 The dice lands on a number less than 7. 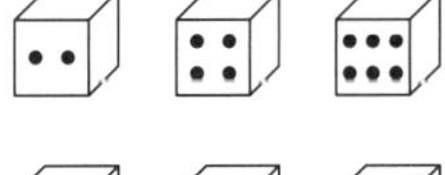

4 The dice lands on a number less than 6. 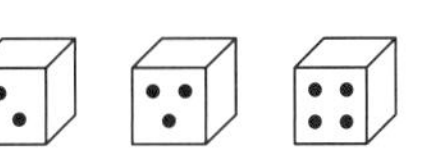 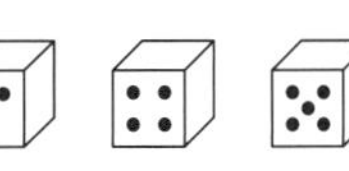

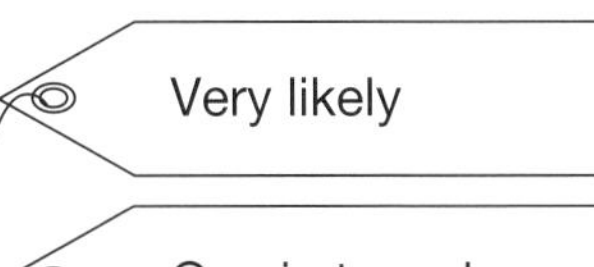

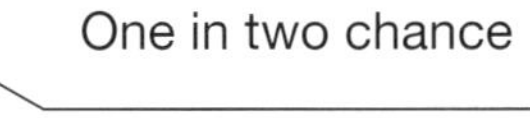

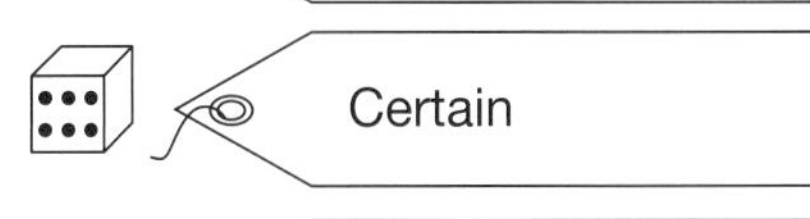

One in six chance

Number and Algebra

SET 1 Basic

1 ☐ + 7 = 15

2 3 + 19

3 2 + 38

4 27 − ☐ = 21

5 23 − 5

6 Product of 6 and 7

7 4 × 6

8 5 × 9

9 6 × 4

10 Sum of 17 and 8

11 9 tens and 6 ones

12 How many tens in 90?

13 Half of 20 then add 7

14 How many tens in 67?

15 Write the largest number you can using 6, 7 and 8.

______

SET 2 Numbers to 10 000

1 Write the number for 10 hundreds.

2 Write the number for 15 tens.

3 Write the number for 4 thousands and 6 ones.

4 What is the place value of 7 in 3476?

5 What is the place value of 6 in 6423?

6 Order these numbers from least to greatest: 1234, 2341, 1341 ______

7 Write the largest number you can using 5, 6, and 3. ______

8 Write the smallest number you can using 9, 4 and 1. ______

9 Write the number ten more than 4501. ______

Mathematical Reasoning

10 Enter the number 6549 on your calculator.

Use subtraction to change the 4 to zero.

What buttons did you press? ______

______

What is your new number? ______

Space Nets

Draw a line to match the box to its net.

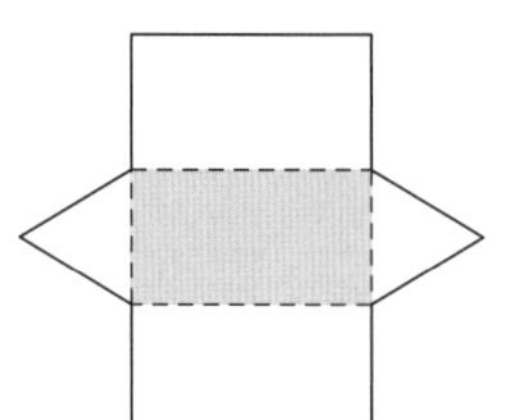

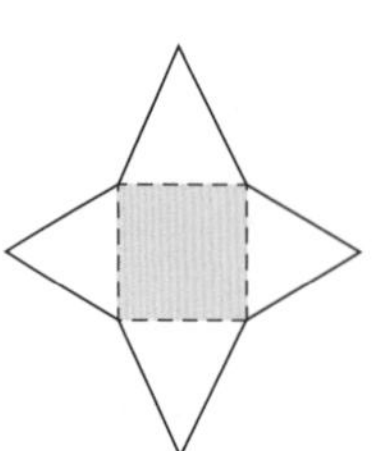

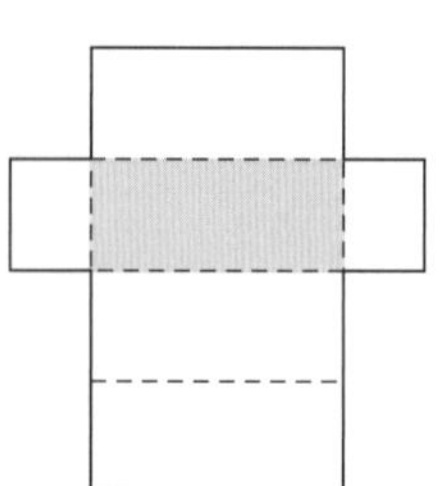

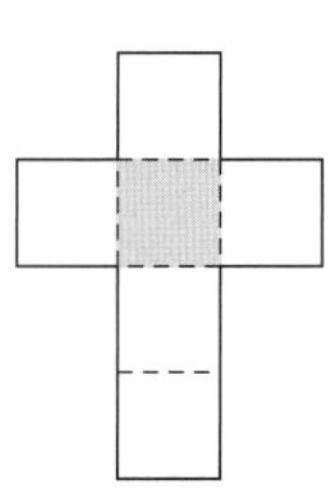

Number and Algebra

SET 3 Input–output machines

Use the rule to complete the input–output machines.

SET 4 Extension

1. Write the numeral for 7 hundreds + 3 tens + 4 ones.
2. Write the numeral for 7 hundreds and 7 ones.
3. 212, 217, 222, [ ], 232
4. How many 5c lollies can be bought for $1.40?
5. 42 + 18 + 43
6. 907c = $ [ ]
7. Share 36 among 5.
8. 7 more than 25
9. How much is $4\frac{1}{2}$ kg of meat at $6 per kg?
10. What is the value of 6 in 764?
11. Are five 50c coins worth more than $2?
12. Subtract 70c from $8.00.
13. What is the value of 70c and 80c?
14. Add 5 tens to 341.
15. Which is heavier, $1\frac{1}{2}$ kg of feathers or 1 kg of metal?
16. Julian saved $376 in November and $561 in December. Does he have enough to buy a bike worth $899?

Measurement Duration – timetables

Big red buses		
Eastwood	9:45	10:05
Denistone	9:48	10:08
Ryde	9:54	10:14
Rhodes	9:59	10:19
Concord	10:06	10:26
Strathfield	10:10	10:30

1. When will the 9:45 bus from Eastwood arrive at Ryde? ________
2. When will the 10:05 bus from Eastwood arrive at Concord? ________
3. How long does it take to travel from Ryde to Rhodes? ________
4. How long does it take to travel from Eastwood to Ryde? ________
5. If I wanted to be at Strathfield at 10:30, what bus should I catch from Denistone? ________

Number and Algebra

SET 1 Basic

1 Add 4 + 34.

2 14 + 7

3 3 + 4 + ☐ = 20

4 Product of 3 and 6

5 3 × 5 + ☐ = 20

6 Multiply 5 and 3.

7 37 – 5

8 Subtract 6 from 25.

9 47 minus 6

10 Sum of 6 and 27

11 Double 21.

12 3 + 3 + ☐ = 16

13 18 plus 7

14 Half of 42

15

How many 10c coins are needed to make $1.90?

☐ 10c coins

SET 2 Equivalent fractions

1 Shade the fractions.

$\frac{1}{2}$

$\frac{3}{4}$

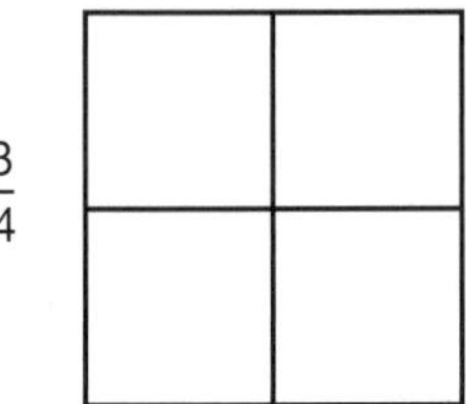

$\frac{8}{10}$

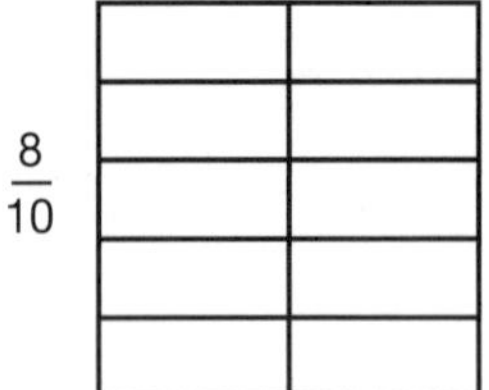

$\frac{4}{5}$

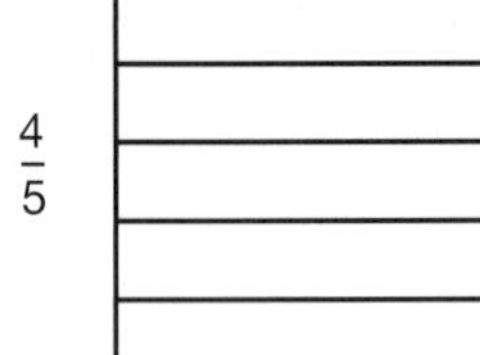

$\frac{4}{8}$

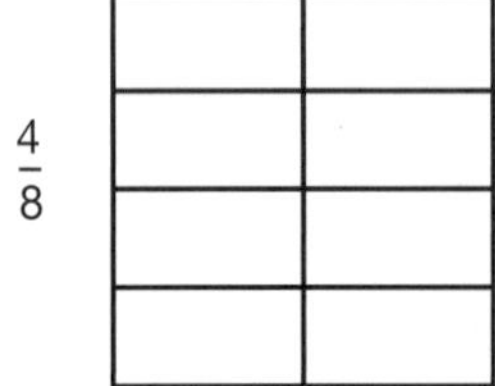

$\frac{6}{8}$

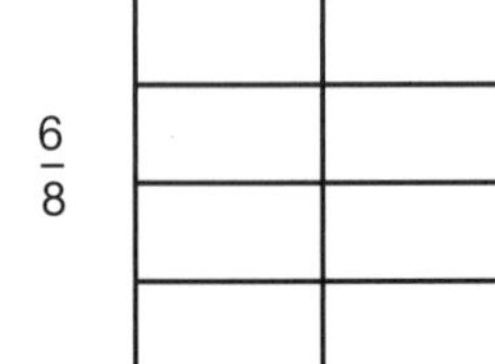

2 Which pairs of fractions shaded above are equivalent?

____

____

Statistics Recording data

James measured the height of four friends. He then recorded each friend's height and age.

Name	Height	Age
Tina	140 cm	8
Zac	143 cm	9
Mohamed	139 cm	11
Simone	145 cm	10

1 Was the oldest also the tallest? ________

2 Was the youngest the shortest? ________

3 Is there a relationship between age and height for his friends? ________

Number and Algebra

SET 3 Numbers beyond 10 000

Write these numbers in expanded form.

1 14 569 = ☐ + ☐ + ☐ + ☐ + ☐

2 11 671 = ☐ + ☐ + ☐ + ☐ + ☐

3 13 811 = ☐ + ☐ + ☐ + ☐ + ☐

4 14 626 = ☐ + ☐ + ☐ + ☐ + ☐

5 13 812 = ☐ + ☐ + ☐ + ☐ + ☐

6 17 613 = ☐ + ☐ + ☐ + ☐ + ☐

Order these numbers from smallest to largest.

7 9671, 8510, 10 811 ____________

8 11 817, 11 781, 11 181 ____________

Order these numbers from largest to smallest.

9 13 671, 18 621, 16 634 ____________

10 15 621, 15 261, 16 121 ____________

SET 4 Extension

1 2 weeks + 6 days = ☐ days

2 What is the value of 8 in 8076?

3 Double 260.

4 How many times can 60 be subtracted from 240?

5 How much will 4 cakes cost at 75c each?

6 What is the value of 3 in 83 921?

7 400 minus 3

8 What number is 7 more than 156?

9 $(3 \times 100) + (2 \times 10) + (6 \times 1)$

10 378c = $ ☐

11 How much will $4\frac{1}{2}$ kg of meat cost at $4 kg?

12 Round 3579 to the nearest 10.

13 How many minutes from 9 am to 2 pm?

14 If lollies are 4 for 20c, how much would 9 cost?

15 How many hours from 3:30 am to 6:30 am?

16 Which number is 100 more than 375?

- ❍ 100
- ❍ 400
- ❍ 475
- ❍ 1375

Measurement Millilitres

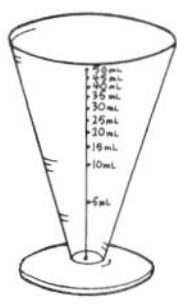

50 mL

20 mL

25 mL

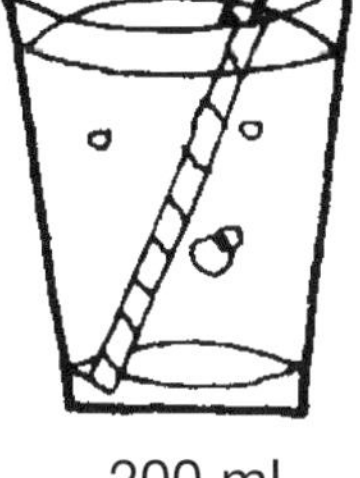

200 mL

Answer the questions.

1 How many bottles of perfume would it take to fill the medicine glass? ____

2 How many medicine glasses would it take to fill the glass? ____

3 How many bottles of White Out would it take to fill the glass? ____

4 How many bottles of perfume would it take to fill the glass? ____

UNIT 33

Number and Algebra

SET 1 Basic

1 19 + 3

2 10 × 9c

3 9 lots of 4

4 ☐ tens + ☐ ones = 57

5 6 × 3

6 Double 8.

7 How many corners does a triangle have?

8 1994 + 5

9 Write 152 in words.

☐

10 300 + 40 + 7

11 $1.72 = ☐ cents.

12 5 × 10c

13 Four squared

14 4 × 9

15 

Write the smallest number you can using 7, 3, 1, 5.

SET 2 Division

Answer the divisions.

1 12 ÷ 2 = ☐

2 15 ÷ 5 = ☐

3 18 ÷ 6 = ☐

4 24 ÷ 2 = ☐

5 20 ÷ 4 = ☐

6 30 ÷ 5 = ☐

7 30 ÷ 6 = ☐

8 $2\overline{)18}$

9 $3\overline{)18}$

10 $4\overline{)24}$

11 $5\overline{)25}$

12 $6\overline{)24}$

13 $2\overline{)16}$

14 $3\overline{)15}$

15 $4\overline{)16}$

16 $5\overline{)35}$

17 $6\overline{)42}$

18 $3\overline{)27}$

Space Maps

Follow the directions to find the secret letter.

1 Start at A.

2 Travel 2 spaces north.

3 Travel 3 spaces east.

4 Travel 1 space south.

5 Travel 3 spaces east.

6 Travel 4 spaces north.

7 Travel 6 spaces west.

8 What letter did you find? ____

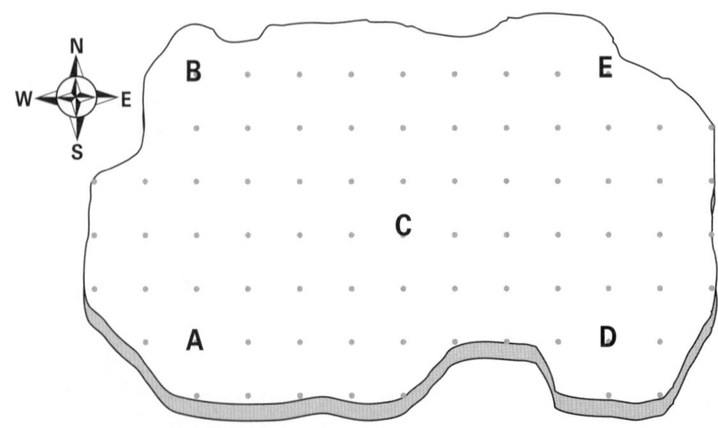

Number and Algebra

SET 3 Estimating

Estimate each number sentence and use a strategy to calculate the correct answer.

	Number sentence	My estimate	Correct answer
1	49 + 84		
2	88 + 144		
3	35 + 192		
4	181 + 156		
5	244 + 362		
6	381 + 281		
7	367 + 211		
8	481 + 468		

Write an estimate then record the correct answer to the problems below.

		Estimate	Correct answer
9	6 × 345		
10	4 × 596		

SET 4 Extension

1 What is the sum of 4, 5 and 8?

2 7 kilograms = ☐ grams

3 Which is larger, 4 × 5 or 20 × 0?

4 Which is larger, 0.3 or $\frac{4}{10}$?

5 What number is 9 more than 1132?

6 How many minutes in $2\frac{1}{2}$ hours?

7 Ten times zero

8 How many grams in $1\frac{1}{4}$ kg?

9 Add 2 tens to 438.

10 What is the product of 4 and 8?

11 Share 21 among 5.

12 How many 50c coins are needed to make $14?

Mathematical Reasoning

13

```
    3 6 7
  + 5 ☐ 2
  -------
    9 4 ☐
```

Explain how you would work out the missing numbers in this question.

____

____

Measurement The gram

List 4 items that have their mass measured in grams.

Item	Mass

UNIT 34

Number and Algebra

SET 1 Basic

1 $8 × 10
2 1999 + 5
3 Sum of 13 and 27
4 Double 15.
5 How many tens in 80?
6 ☐ + 8 = 37
7 5^2
8 Double 8 and double again.
9 Product of 5 and 7
10 Sum of 32 and 15
11 Six books at $5 each
12 $5 × 3
13

Write the smallest number you can using 5, 9, 6, 5.

____

SET 2 Adding 3 addends

Answer the additions.

1

	HUND	TENS	ONES
	3	4	3
	1	1	3
+	1	3	7

2

	HUND	TENS	ONES
	3	2	6
		2	6
+	1	4	6

3

	HUND	TENS	ONES
	1	0	7
	2	8	0
+	1	0	5

4

	HUND	TENS	ONES
	4	5	7
	3	4	3
+		3	7

5

	HUND	TENS	ONES
	7	6	5
	1	3	4
+		4	9

6

	HUND	TENS	ONES
	2	3	6
		8	2
+			8

Complete these magic squares.

7

9	4	5
		10
7		3

8

10		6
	7	11
8		4

9

	4	11
10	8	6
	12	

10

14	9	10
		15
12		8

Statistics Column graphs

Favourite Pets

Cat	𝍸	Bird	||||
Dog	𝍸 |		
Fish	||||		

Count the tally marks and write how many children had a:

1 Cat for a pet? ☐
2 Bird for a pet? ☐
3 Fish for a pet? ☐
4 Dog for a pet? ☐

Favourite Pets

Cat	Dog	Fish	Bird

5 Colour the graph to represent the data.
6 How many children were surveyed altogether? ☐

Number and Algebra

SET 3 Grid references

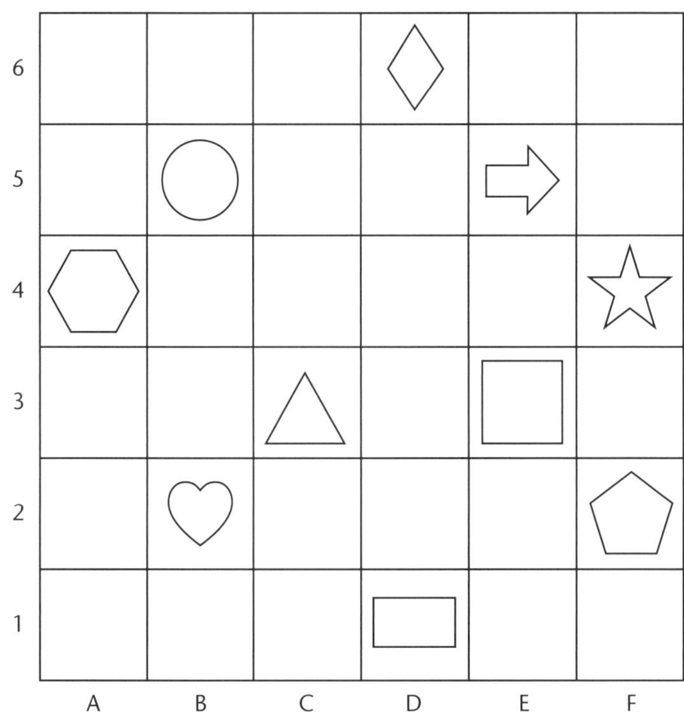

What can be found at the grid references?

1 C3 _______
2 B5 _______
3 D1 _______
4 E5 _______
5 A4 _______
6 B2 _______
7 F4 _______
8 D6 _______
9 F2 _______
10 E3 _______

SET 4 Extension

1 Share 30c between 5 people.
2 How many 10c coins in $3.00?
3 15 + 4 + 3
4 What is the value of 3 in 2377?
5 $10 – $1.50
6 Write 267 in words.
7 How much is $\frac{1}{4}$ of 12 books?
8 What number is 8 more than 1328?
9 How many 20c coins in $4?
10 Which is larger, $\frac{1}{4}$ or $\frac{1}{3}$?
11 297, 300, 303, ____, ____
12 What number is 20 less than 468?
13 Share 27 among 4.
14 How much are 4 books at $6 each?
15 How much are 7 screwdrivers at $4 each?
16 If 6 books cost $24, how much do 5 books cost?

Measurement Recognising metric units

Choose the correct measuring units to measure the items.

	Items	g	kg	L	mL	m	cm
1	The length of a basketball court						
2	The length of a paperclip						
3	The mass of a banana						
4	The length of a glue stick						
5	The capacity of a watering can						
6	The capacity of a medicine glass						
7	The length of a bus						

UNIT 35

Number and Algebra

SET 1 Basic

1 6 + 4 + 7

2 7 + 3 + 9

3 18 – 6

4 17 – 8

5 4 × 10

6 5 × 6

7 7 × 3

8 8 × 2

9 Add 27 to 34.

10 Multiply 3 by 5.

11 (6 + 4) × 3

12 How many minutes in 2 hours?

13 Difference between 42 and 20

14 42 – ☐ = 30

15

SET 2 Equivalent number sentences

Write another number sentence to balance the ones given. The first one is done for you.

1 30 + 15 = 50 – 5

2 5 × 2 × 2 = ☐

3 27 + 23 = ☐

4 30 – 6 – 7 = ☐

5 50 – 15 = ☐

6 45 + 45 = ☐

7 25 ÷ 5 = ☐

8 100 – 30 = ☐

9 5 × 3 × 2 = ☐

Space Giving directions

Complete the instructions to describe the path drawn on the map. It has been started for you.

1 Start at ■. Travel east 4 spaces.

2 Travel north 3 spaces.

3 ______________________________

4 ______________________________

5 ______________________________

6 ______________________________

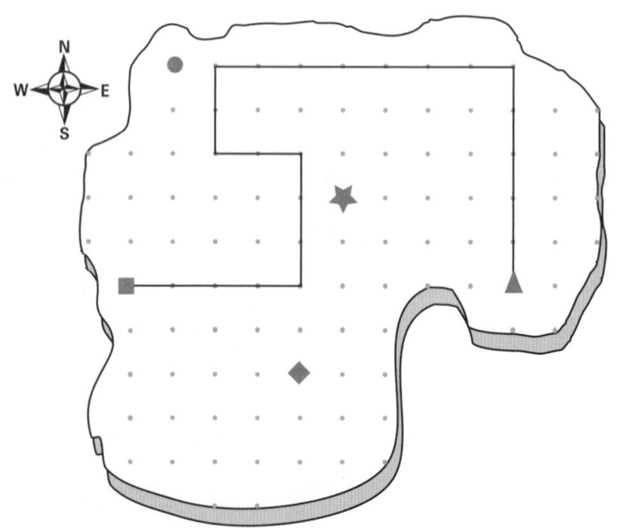

Number and Algebra

SET 3 2-digit multiplication

Complete the multiplications.

1
$$\begin{array}{r} 27 \\ \times\ \ 2 \\ \hline \\ \hline \end{array}$$

2
$$\begin{array}{r} 58 \\ \times\ \ 3 \\ \hline \\ \hline \end{array}$$

3
$$\begin{array}{r} 45 \\ \times\ \ 3 \\ \hline \\ \hline \end{array}$$

4
$$\begin{array}{r} 75 \\ \times\ \ 5 \\ \hline \\ \hline \end{array}$$

5
$$\begin{array}{r} 39 \\ \times\ \ 3 \\ \hline \\ \hline \end{array}$$

6
$$\begin{array}{r} 62 \\ \times\ \ 5 \\ \hline \\ \hline \end{array}$$

7 Trent bought 8 books at $5 each. He paid for the books with a $100 note. How much change did he receive? 

8 Jesse bought 5 books at $7 each. She paid for the books with a $50 note. How much change did she receive? 

SET 4 Extension

1 Write the largest number you can using 2, 8, 4.

2 How many sides on 7 pentagons?

3 If 6 pens cost $30, how much do 4 cost?

4 What is half of 48?

5 How many legs on 3 spiders and 2 dogs?

6 20 less than 496

7 20 more than 507

8 Value of 4 in 428

9 $(2 \times 10) + 7$

10 $2 \times (10 - 7)$

11 How much are 5 books at $8 each?

12 $27 + \square = 40$

13 Write 797 in words.

____

14 420 take away 200

15 Subtract 40 from 280.

16 Estimate an answer to 39 + 141.

17 $\frac{1}{3}$ of $15

Measurement Length using decimal notation

Draw a line to match each item to an appropriate length.

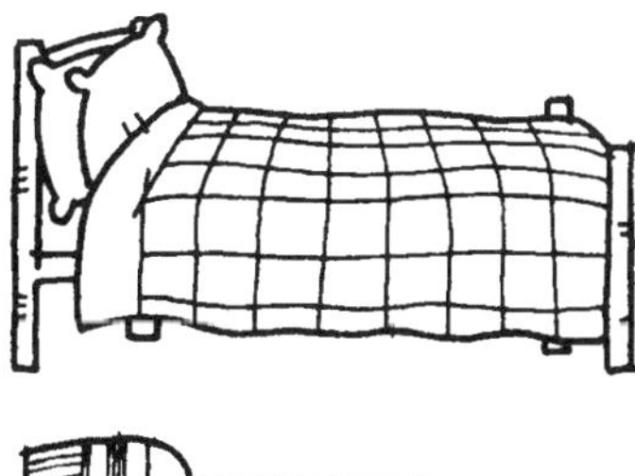

4 m
1.95 m
1.56 m
0.15 m

Maths helpers

Length

10 millimetres (mm) = 1 centimetre (cm)

100 centimetres (cm) = 1 metre (m)

1000 metres (m) = 1 kilometre (km)

Mass

1000 grams (g) = 1 kilogram (kg)

1000 kg = 1 tonne (t)

Capacity

1000 millilitres (mL) = 1 litre (L)

Time

60 seconds = 1 minute

60 minutes = 1 hour

24 hours = 1 day

7 days = 1 week

14 days = 1 fortnight

12 months = 1 year

52 weeks = 1 year

365 days = 1 year

366 days = 1 leap year

10 years = 1 decade

100 years = 1 century

Months of the year

Thirty days has September, April, June and November. All the rest have thirty-one, except February alone, which has twenty-eight days clear and twenty-nine days each leap year.

Seasons

Summer: December, January, February

Autumn: March, April, May

Winter: June, July, August

Spring: September, October, November

Roman numerals

1 = I

2 = II

3 = III

4 = IV

5 = V

6 = VI

7 = VII

8 = VIII

9 = IX

10 = X

20 = XX

30 = XXX

40 = XL

50 = L

60 = LX

70 = LXX

80 = LXXX

90 = XC

100 = C

500 = D

1000 = M

Multiplication facts

×	0	1	2	3	4	5	6	7	8	9	10
0	0	0	0	0	0	0	0	0	0	0	0
1	0	1	2	3	4	5	6	7	8	9	10
2	0	2	4	6	8	10	12	14	16	18	20
3	0	3	6	9	12	15	18	21	24	27	30
4	0	4	8	12	16	20	24	28	32	36	40
5	0	5	10	15	20	25	30	35	40	45	50
6	0	6	12	18	24	30	36	42	48	54	60
7	0	7	14	21	28	35	42	49	56	63	70
8	0	8	16	24	32	40	48	56	64	72	80
9	0	9	18	27	36	45	54	63	72	81	90
10	0	10	20	30	40	50	60	70	80	90	100

Addition facts

+	2	3	4	5	6	7	8	9	10	11	12
2	4	5	6	7	8	9	10	11	12	13	14
3	5	6	7	8	9	10	11	12	13	14	15
4	6	7	8	9	10	11	12	13	14	15	16
5	7	8	9	10	11	12	13	14	15	16	17
6	8	9	10	11	12	13	14	15	16	17	18
7	9	10	11	12	13	14	15	16	17	18	19
8	10	11	12	13	14	15	16	17	18	19	20
9	11	12	13	14	15	16	17	18	19	20	21
10	12	13	14	15	16	17	18	19	20	21	22
11	13	14	15	16	17	18	19	20	21	22	23
12	14	15	16	17	18	19	20	21	22	23	24

UNIT 1 Number and Algebra

SET 1

1 8
2 10
3 4
4 9
5 8
6 11
7 19
8 8
9 5
10 6
11 0
12 8
13 8
14 7
15 5

SET 2

1 9
2 17
3 11
4 16
5 18
6 13
7 16
8 13
9 6
10 $20
11 16 cm
12 19
13 9
14 $14
15 36 + 12 = 48

SET 3

1 12, 15, 18
2 20, 22, 24
3 80, 90, 100
4 20, 25, 30
5 22, 26, 30
6 32, 35, 38
7 50, 55, 60
8 70, 75, 80 (Add 5.)
9 28, 26, 24 (Subtract 2.)
10 41, 46, 51 (Add 5.)

SET 4

1 5 tens + 5 ones
2 17 + 6
3 $24
4 56
5 Fifteen
6 $38
7 20 + 7 or 14 + 13 or 15 + 12
8 15 + 12 or 14 + 13
9–12 (No fixed order) 18 + 2, 17 + 3, 16 + 4, 15 + 5

Space

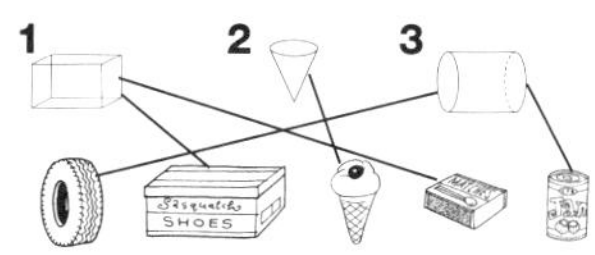

Measurement

Hands on

UNIT 2 Number and Algebra

SET 1

1 9
2 10
3 11
4 11
5 13
6 11
7 10
8 5
9 3
10 3
11 9
12 5
13 12
14 3
15 19

SET 2

1 10
2 12
3 7
4 $9
5 $12
6 3
7 7 km
8 50 cm
9 14
10 4
11 9 min
12 6
13 5
14 11
15 4
16 16

SET 3

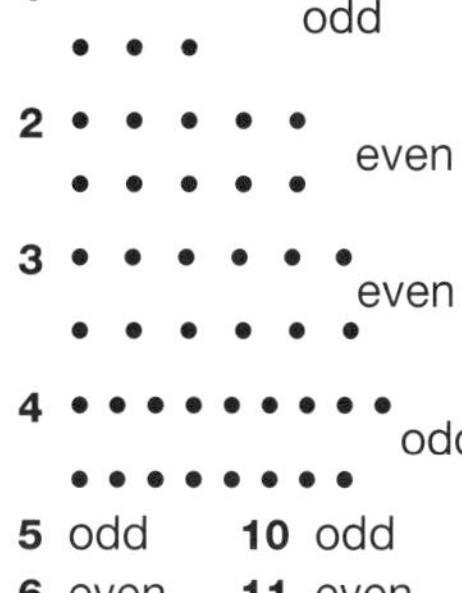

5 odd
6 even
7 odd
8 even
9 even
10 odd
11 even
12 odd
13 even

SET 4

1 4
2 42
3 9
4 12
5 Summer
6 31
7 25 December
8 5 June
9 20
10 67
11 3
12 19
13 10 km
14 20
15 Hands on

Number and Algebra

1 3 + 3 = 6
2 3 + 3 + 3 + 3 = 12
3 4 + 4 + 4 + 4 = 16
4 4 + 4 + 4 + 4 + 4 = 20

Statistics

1 5
2 pizza
3 pie
4 3
5 7
6 sushi and burger

UNIT 3 Number and Algebra

SET 1

1 11
2 7
3 11
4 9
5 6
6 13
7 4
8 1
9 0
10 6
11 4
12 1
13 6
14 14
15 22

SET 2

1 $2 \times 2 = 4$
2 $2 \times 4 = 8$
3 $3 \times 4 = 12$
4 $4 \times 5 = 20$
5 $4 \times 6 = 24$

SET 3

1 5 ✓
2 11 ✗
3 7 ✓
4 8 ✓
5 12 ✗
6 12 − 7 = 5, 5 + 7 = 12
7 18 − 5 = 13, 13 + 5 = 18
8 38 − 25 = 13, 13 + 25 = 38

SET 4

1 6 tens, 9 ones
2 17 + 9
3 51
4 67
5 16
6 1 ten, 9 ones
7 9 tens
8 54
9 41
10 Autumn
11 47
12 3 tens, 6 ones
13 60, 70
14 38
15 $6 \times 4 = 24$ or $8 \times 3 = 24$
16 $2 \times 12 = 24$ or $6 \times 4 = 24$

Space

10 CAT AND DOG

Statistics

1 10
2 13
3 3
4 Kia
5 Toyota and Mazda

Answers

UNIT 4 Number and Algebra

SET 1

1 14
2 13
3 14
4 13
5 14
6 17
7 15
8 14
9 9
10 11
11 12
12 2
13 16
14 8
15 $12

SET 2

1 13, 7
2 14, 6
3 19, 9
4 17, 5
5 16, 13
6 20, 2
7 18, 9
8 11
9 7
10 $28

SET 3

1 2
2 4
3 6
4 8
5 10
6 12
7 14
8 16
9 18
10 20
11 4 arrays/ Hands on e.g.
4 × 5
5 × 4
10 × 2
2 × 10

SET 4

1 75
2 5
3 11
4 14
5 132
6 20c
7 5 each
8 27
9 8 tens, 1 one
10 10
11 80
12 176
13 Ninety-two
14 24
15 May
16 Rectangular prism

Space

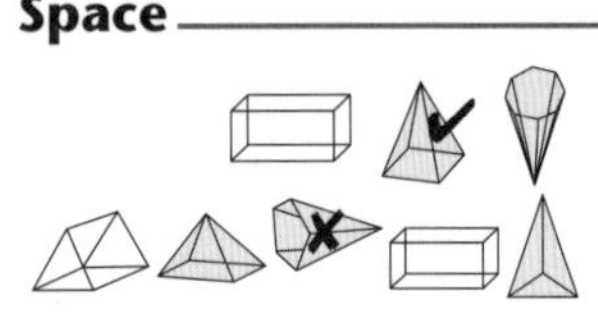

Measurement

1 10 cm
2 12 cm
3 5 cm
4 9 cm
5 13 cm

UNIT 5 Number and Algebra

SET 1

1 11
2 12
3 $12
4 1
5 10
6 12
7 14
8 24
9 7
10 20
11 20
12 40
13 8
14 20
15 20

SET 2

1 13
2 130
3 17
4 170
5 8
6 80
7 11
8 110
9 15
10 150
11 110, 130, 130
150, 120, 140
12 50, 70, 90
110, 1300, 1600

SET 3

1 13, 18, 23, 28, 33, 38
2 21, 31, 41, 51, 61, 71
3 ✓
4 ✓
5 ✗
6 ✓

SET 4

1 5
2 9 tens, 3 ones
3 4
4 40c
5 Twelve
6 30
7 3
8 September
9 18
10 40
11 67
12 8
13 No
14 45
15 6
16 Octagon

Number and Algebra

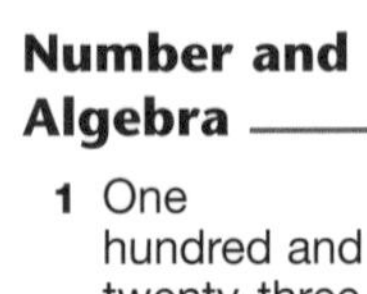

1 One hundred and twenty-three
2 Three hundred and sixty-seven
3 237
4 460
5 370, 360
6 463
7 274

Space

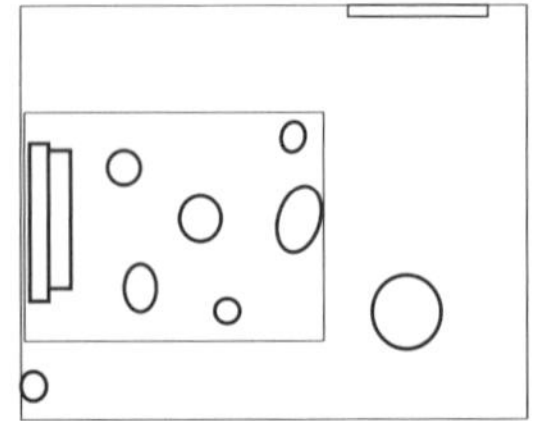

UNIT 6 Number and Algebra

SET 1

1 1
2 4
3 9
4 7
5 9
6 9
7 12
8 4
9 6
10 3
11 14
12 11
13 15c
14 0
15 $20

SET 2

1 500 + 30 + 9
2 600 + 30 + 5
3 200 + 70 + 8
4 300 + 80 + 6
5 700 + 90 + 4
6 400 + 90 + 7
7 <
8 <
9 >
10 <
11 >

SET 3

1 2
2 5
3 4
4 6
5 3
6 5
7 25
8 40
9 50
10 45
11 30 sheep
12 $40
13 $35

SET 4

1 2
2 31
3 6
4 5
5 5c
6 Spring
7 60
8 14
9 20c
10 75
11 43
12 180
13 About 60
14 4 each
15 18, 20, 22, 24, 26, 28
16 24, 26, 28, 30, 32, 34

Numbers to 5000

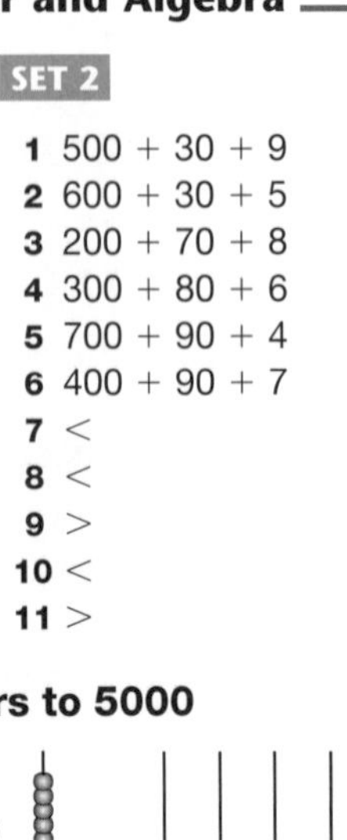

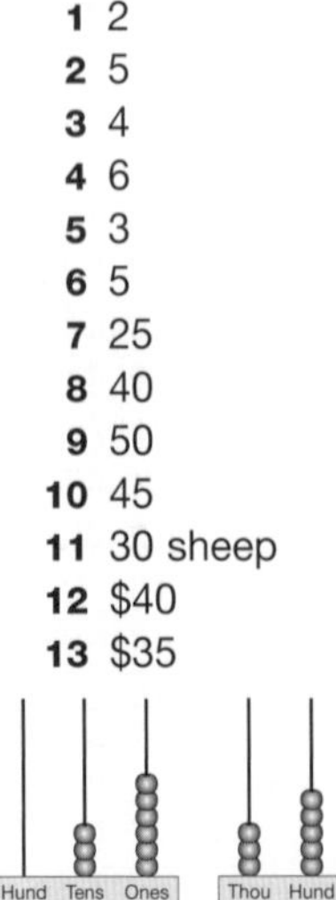

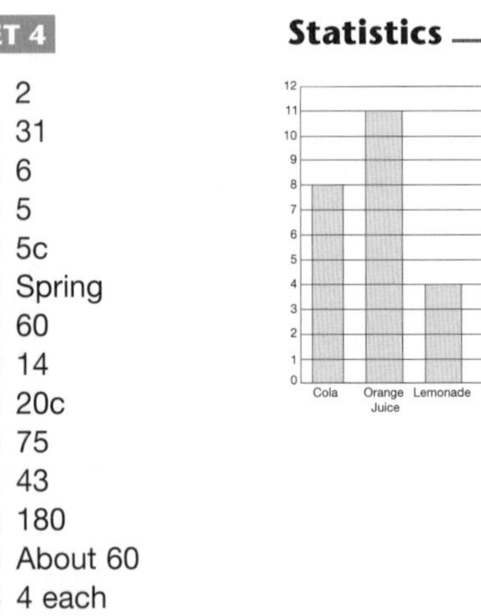

Statistics

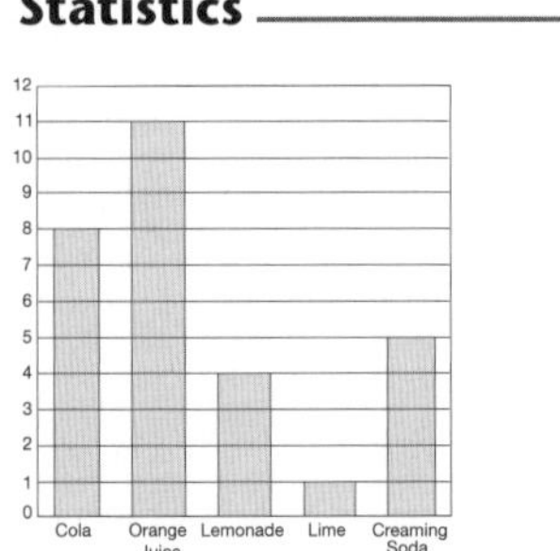

UNIT 7 Number and Algebra

SET 1

1 9
2 9
3 5
4 10
5 10
6 4
7 3
8 3
9 6
10 14
11 2, 4, 6, 8, 10, 12
12 14
13 14
14 9
15 9

SET 2

1 3
2 4
3 6
4 2
5 8
6 4
7 16
8 2
9 1

SET 3

1 19
2 25
3 31
4 24
5 38
6 24
7 46
8 93
9 6
10 $20
11 31 cm
12 85
13 35
14 40

SET 4

1 $5
2 81
3 64
4 16
5 Sixteen
6 No
7 69
8 24
9 9
10 May
11 2 3 4 5 6 7 8

Measurement

1 bucket
2 glass or 1 litre milk

Number and Algebra

1 30
2 50
3 40
4 70
5 0
6 40
7 80

8

	3	5	6	2	9	8	7	4
X3	9	15	18	6	27	24	21	12

9

	3	5	6	2	9	8	7	4
X2	6	10	12	4	18	16	14	8

10

	3	5	6	2	9	8	7	4
X10	30	50	60	20	90	80	70	40

UNIT 8 Number and Algebra

SET 1

1 6
2 8
3 11
4 9
5 1
6 4
7 6
8 7
9 0
10 28
11 32
12 20
13 50
14 11
15 50c

SET 2

1 5
2 50
3 4
4 40
5 7
6 70
7 4
8 40
9 400
10 200
11 30 kg
12 $500

SET 3

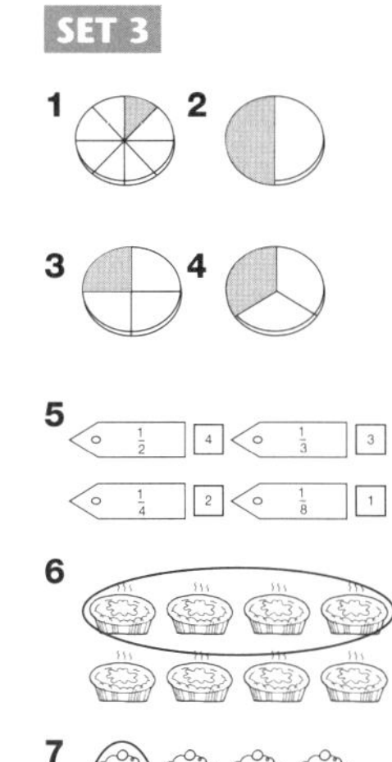

SET 4

1 12
2 15
3 6
4 March
5 190
6 Seventeen
7 14
8 86
9 41
10 31
11 30
12 2
13 About 70
14 Hands on

Space

a squares
b rectangles, squares
c triangles, squares, rectangles
d squares, triangles

Number and Algebra

1 60
2 94
3 65
4 65
5 71
6 70

UNIT 9 Number and Algebra

SET 1

1 3
2 6
3 9
4 9
5 9
6 9
7 12
8 18
9 5c
10 1
11 15
12 10
13 12
14 24
15 15

SET 2

1 27 + 32 becomes 50 + 9 = 59
2 44 + 23 becomes 60 + 7 = 67
3 35 + 17 becomes 40 + 12 = 52
4 58 + 22 becomes 70 + 10 = 80
5 45
6 33
7 59
8 48
9 51
10 68

SET 3

1 83
2 74
3 83
4 82
5 76
6 93
7 85
8 74

SET 4

1 2
2 28
3 80
4 30
5 Twenty-three
6 4
7 February
8 24
9 95
10 5
11 8 tens, 5 ones
12 4
13 August
14 20
15 2 20 6 3 4 12 11 29 Hands on One example

Statistics

1 5
2 9
3 2
4 29
5 blond and black
6 Hands on

Measurement

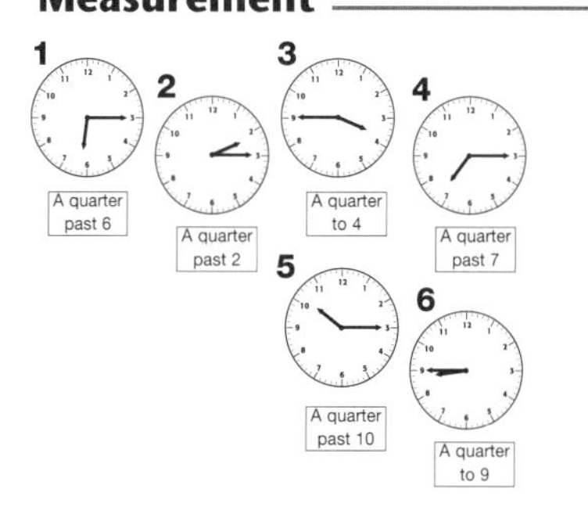

Answers

UNIT 10 Number and Algebra

SET 1

1 13
2 $2
3 12
4 12
5 10
6 24
7 16
8 5
9 13
10 16
11 95c
12 6
13 30
14 20
15 $14

SET 2

1 17 – 6 = 11
6 + 11 = 17
2 23 – 9 = 14
9 + 14 = 23
3 35 – 31 = 4
31 + 4 = 35
4 57 – 52 = 5
52 + 5 = 57
5 27 – 18 = 9
18 + 9 = 27
6 45 – 32 = 13
32 + 13 = 45
7 40 + 20 + 10 = 70
30 + 30 + 10 = 70
30 + 20 + 20 = 70

SET 3

1

4	6	8	10	12	14	16	18

Rule – Count on 2 or add 2

2

3	6	9	12	15	18	21	24

Rule – Count on 3 or add 3

3

15	20	25	30	35	40	45	50

Rule – Count on 5 or add 5

4
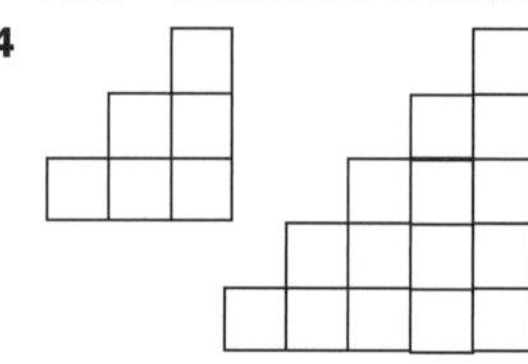

5 21

SET 4

1 8 tens, 2 ones
2 3 + 8
3 30
4 90c
5 18
6 2 February
7 35
8 10
9 2 r 4
10 Ninety-five
11 62
12 2 hund, 3 tens, 4 ones
13 18
14 Twenty-eight
15 93

Space

Angles 1 and 3 are blue.
Angles 2, 4 and 5 are red.

Measurement

	Item	Longer than 1m	About the same as 1 m	Shorter than 1m
1	broom	■		
2	chair height		■	■
3	refrigerator	■		
4	dustbin			■
5	shovel		■	
6	telephone booth	■		

UNIT 11 Number and Algebra

SET 1

1 10
2 10
3 20
4 20
5 3
6 4
7 5
8 6
9 12
10 24
11 16
12 20
13 14
14 15
15 $12

SET 2

1 14, 24, 34, 44, 54, 64, 74, 84
2 130, 140, 150, 160, 170, 180, 190, 200
3 57, 67, 77, 87, 97, 107, 117, 127
4 93, 83, 73, 63, 53, 43, 33, 23
5 100, 90, 80, 70, 60, 50, 40, 30
6 255, 245, 235, 225, 215, 205, 195, 185
7 34, 44, 54, 64, 74
8 122, 132, 142, 152, 162
9 54, 64, 74, 84, 94
10 125, 135, 145, 155 165
11 150, 160, 170, 180, 190

SET 3

1 Any 1 part
2 Any 3 parts
3 Any 2 parts
4 Any 1 part
5 Any 1 part
6 Any 2 parts
7 Any 3 squares
8 Any 2 triangles
9 Any 2 circles
10 Yes

SET 4

1 16
2 11, 12, 17, 19
3 58
4 One hundred and eighty-seven
5 325
6 175c
7 15
8 $1.50
9 21
10 200
11 46
12 50c
13 13
14 3
15 5
16 Pentagon

Number and Algebra

1 28
2 37
3 26
4 55
5 39
6 15

Statistics

1 Apple
2 Pear
3 Orange and banana
4 Apple

UNIT 12 Number and Algebra

SET 1

1 20
2 17
3 18
4 16
5 13
6 11
7 17
8 18
9 18
10 36
11 10
12 14
13 4
14 10
15 $15

SET 2

1 0, 2, 4, 18, 8, 16, 12, 24
2 20, 80, 60, 140, 120, 40, 44, 160
3 39
4 82
5 61
6 53
7 Hands on (some examples below)
$5 + $5 + $5 +
$5 + $5 + $5 +
$5 + $5 +
or
$7 + $6 + $7 +
$6 + $4 + $5 + $5

SET 3

1 2, 4, 8, 16, 32, 64
2 3, 6, 12, 24, 48, 96
3 5, 10, 20, 40, 80, 160
4 128, 64, 32, 16, 8, 4
5 544, 272, 136, 68, 34, 17
6 608, 304, 152, 76, 38, 19
7 448, 224, 112, 56, 28, 14

SET 4

1 Summer
2 1, 2, 3, 5, 7
3 13
4 9
5 58
6 Sixty-four
7 89
8 12
9 15
10 $24
11 28
12 100
13 120
14 25
15 114

Space

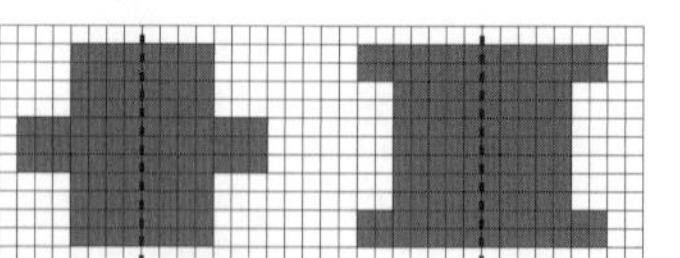

Statistics

UNIT 13 Number and Algebra

SET 1

1 9
2 11
3 16
4 $4
5 19
6 18
7 20
8 21
9 13
10 5
11 12
12 14
13 20
14 6
15 8

SET 2

1 55
2 55
3 56
4 28
5 58
6 48
7 38
8 58

SET 3

1 Any 1 part
2 Any 1 part
3 Any 2 parts
4 Any 1 part
5 Any 1 circle
6 Any 2 squares
7 Any 3 triangles

SET 4

1 82
2 73
3 3
4 15
5 $19
6 37
7 22
8 24
9 3 each
10 20
11 August
12 65c
13 60
14 20
15 September
16

Space

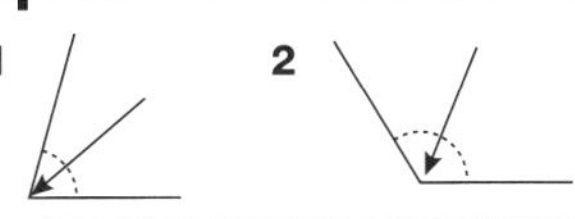

3

Angle	Smaller than	Right angle	Larger than

Measurement

1 8 L
2 3 L
3 2 L
4 1 L
5 5
6 2

UNIT 14 Number and Algebra

SET 1

1 10
2 11
3 14
4 9
5 5
6 12
7 22
8 17
9 15
10 19
11 14
12 Yes
13 7
14 20
15 7th

SET 2

1 6
2 4
3 12
4 2
5 3
6 8
7 5
8 3
9 5
10 5
11 5

SET 3

1 60
2 80
3 90
4 80
5 90
6 90
7 60 km
8 80 km
9 80 km

SET 4

1 213
2 30
3 406
4 25
5 25
6 24
7 2 hund, 7 tens, 6 ones
8–11 Hands on
12 13

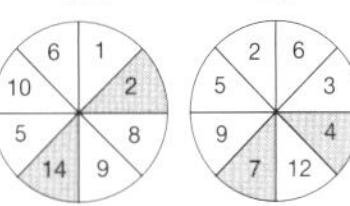

Space

a b c d

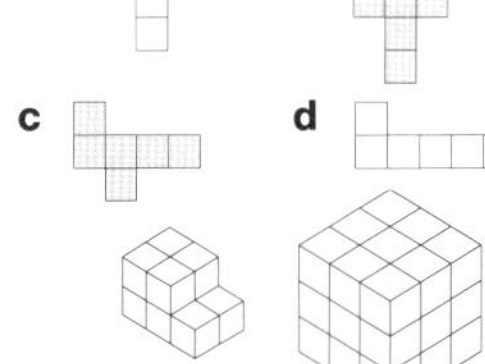

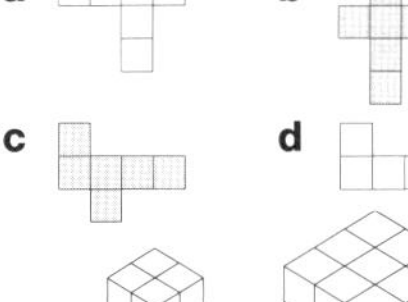

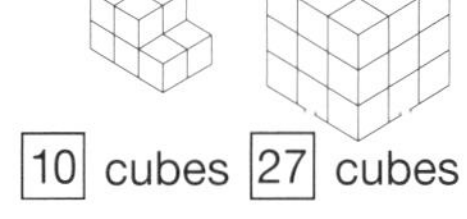

10 cubes 27 cubes

Statistics

1 John, Tim, Kelly
2 Jessica, Con, Abdul, Sam
3 Kim, Greg, Maria, Soula

UNIT 15 Number and Algebra

SET 1

1 13
2 15
3 14
4 23
5 10
6 9
7 8
8 12
9 20
10 14
11 12
12 12
13 325
14 13
15 12

SET 2

Hands on
Some possible answers

		10	20		
1	I	II	III	II	
2	I	I	I	I	I
3	I	I	IIIIII	I	
4	I	I	I	III	
5	III		I	I	I
6	III	III	II	II	

7 $1.80
8 $2.00
9 $17
10 22

SET 3

1 3674

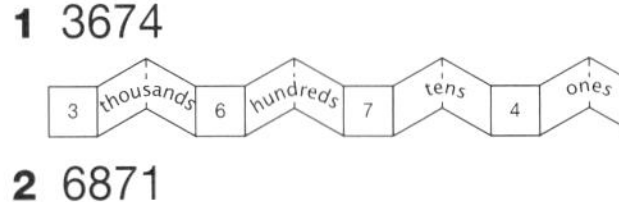

2 6871

6 thousands 8 hundreds 7 tens 1 ones

3 8150

8 thousands 1 hundreds 5 tens 0 ones

4 700
5 70
6 500
7 3000
8 9
9 3000

SET 4

1 45
2 24
3 21
4 About 110
5 360
6 15
7 25
8 327
9 8
10 300 cm
11 One hundred and seventy-two
12 40
13 2
14 $1 or 100c
15 14
16 15

Space

1–5

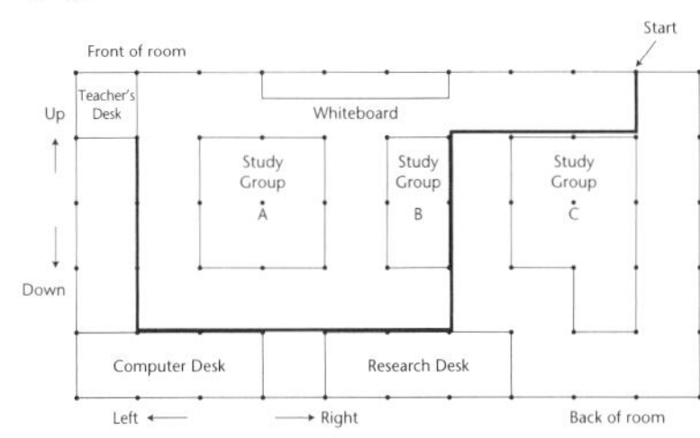

6 Teacher's Desk

Measurement

1 2 3

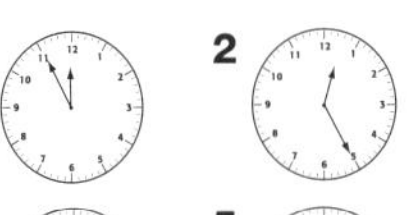

4 5

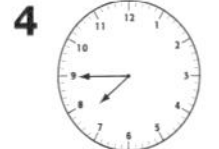

Answers

UNIT 16 Number and Algebra

SET 1

1 23
2 16
3 50
4 20
5 22
6 20
7 20
8 14
9 20
10 15
11 15
12 20
13 4
14 2
15 $20

SET 2

1

9	4	5
2	6	10
7	8	3

2

7	2	9
8	6	4
3	10	5

3

9	4	11
10	8	6
5	12	7

4

14	9	10
7	11	15
12	13	8

SET 3

1 6
2 10
3 14
4 18
5 12
6 30
7 50
8 70
9 90
10 60
11 15
12 25
13 35
14 45
15 30
16 40
17 Hands on

1×24, 24×1
2×12, 12×2
3×8, 8×3
4×6, 6×4

SET 4

1 1300
2 56
3 9 tens, 5 ones
4 1 hund, 6 tens, 2 ones
5 627
6 10 each
7 135
8 Thirty-nine
9 15
10 150, 200, 250, 300, 350
11 $2.50
12 239
13 7th
14 3
15 $1.20

Statistics

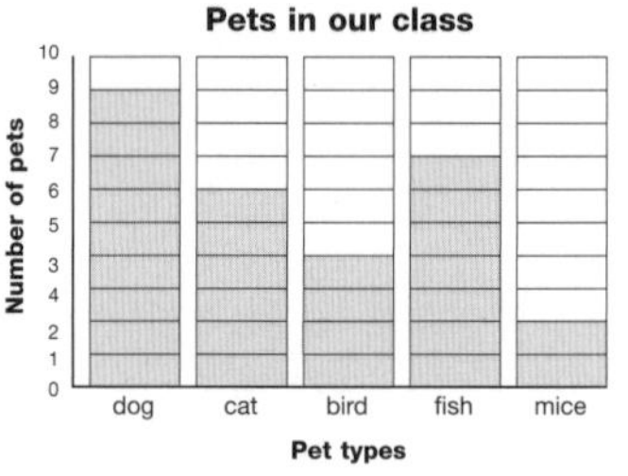

Measurement

Scale A and D are correct.

UNIT 17 Number and Algebra

SET 1

1 4
2 16
3 12
4 4
5 60
6 50
7 3
8 10
9 19
10 Friday
11 26
12 30, 40
13 $3, $6, $9, $12, $15

SET 2

1 9 + 1 + 7 = 17
2 18 + 2 + 5 = 25
3 48 + 2 + 7 = 57
4 64 + 6 + 2 = 72
5 36 + 4 + 1 = 41
6 45 + 5 + 2 = 52
7 98 + 2 + 30 = 130
8 96 + 4 + 40 = 140
9 95 + 5 + 30 = 130
10 86 + 14 + 20 = 120
11 185 + 15 + 30 = 230
12 176 + 24 + 30 = 230

SET 3

1 $4
2 3
3

24	32	16	12	8	40	4
6	8	4	3	2	10	1

4 5 for $30 at $6 each

SET 4

1 17
2 1000
3 2 dozen
4 754
5 2300
6 30
7 0
8 About 280
9 89
10 12
11 5 each
12 155, 161
13 $1.50
14 150 cm
15 7 June

Number and Algebra

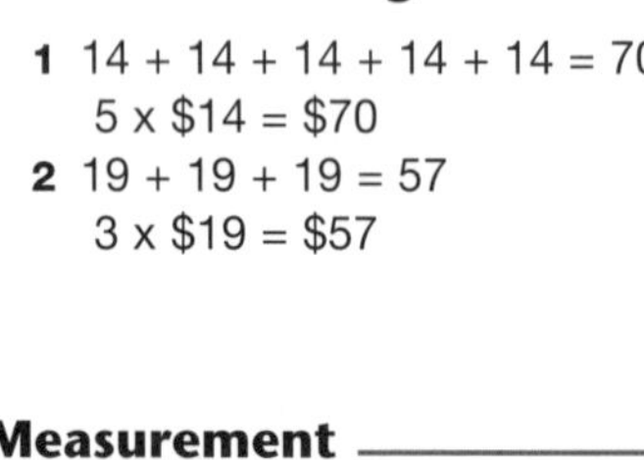

1 14 + 14 + 14 + 14 + 14 = 70
5 x $14 = $70
2 19 + 19 + 19 = 57
3 x $19 = $57

Measurement

1 7 cm
2 3 cm
3

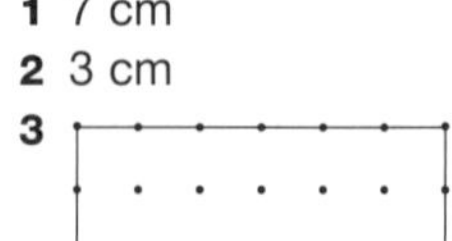

UNIT 18 Number and Algebra

SET 1

1 8
2 18
3 28
4 38
5 7
6 11
7 10
8 8
9 6
10 28
11 45
12 32
13 12
14 35
15 $28

SET 2

1 4
2 2
3 12
4 4
5 20
6 40
7 8
8 16
9 20
10 32
11 Hands on

SET 3

1 $75, $70, $65, $60, $55, $50
2 $16, $20, $24, $28, $32, $36
3 10, 13, 16, 19, 22, 25, 28, 31
4 62, 58, 54, 50, 46, 42, 38, 34
5 0, 9, 18, 27, 36, 45, 54, 63
6 57, 65, 73, 81, 89, 97, 105, 113
7 20, 30, 40, 50, 60, 70, 80, 90
Rule: Count by 10 or add 10
8 9, 12, 15, 18, 21, 24, 27, 30
Rule: Count by 3 or add 3

SET 4

1 8
2 3
3 3
4 24
5 $35
6 473
7 971
8 179
9 976
10 $1
11 16 + 8 + 4 + 2 + 1 = 31

Number and Algebra

1 $11 + $27 = $38
2 $12 + $57 = $69

Space

Hands on. Some examples below.

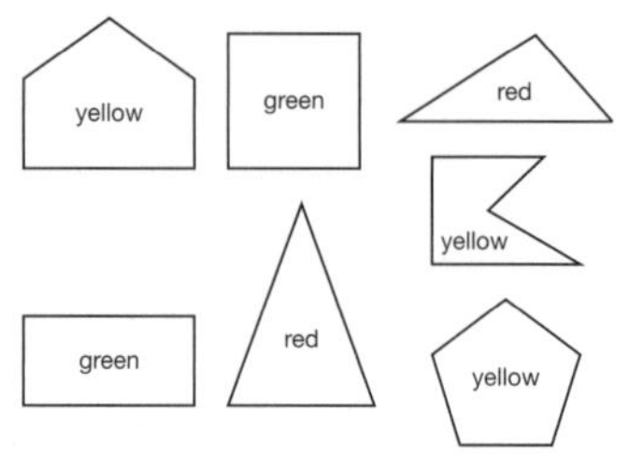

UNIT 19 Number and Algebra

SET 1

1 13
2 $4
3 8
4 $16
5 5
6 20
7 12
8 11
9 3
10 4
11 24
12 19
13 35
14 90
15 7

SET 2

1 6 thousands 3 hundreds 7 tens 6 ones
2 5 thousands 5 hundreds 9 tens 1 ones
3 8 thousands 3 hundreds 8 tens 5 ones
4 9 thousands 2 hundreds 9 tens 6 ones
5 2000 + 500 + 70 + 9
6 3000 + 600 + 20 + 8
7 5000 + 100 + 90 + 6
8 7000 + 800 + 90 + 6
9 9000 + 400 + 90 + 4
10 7341, 7437, 8471

SET 3

1	**2**	**3**
20	9	2
30	21	4
25	18	8
35	24	18
45	27	12
40	15	10

4 9
5 18
6 12
7 15
8 21
9 24
10 30
11 45
Winning card is b

SET 4

1 $16
2 35
3 51
4 240
5 12 each
6 $95
7 675
8 349
9 7
10 223, 229, 235, 241, 247
11 4 5 6   4 6 5
5 4 6   5 6 4
6 4 5   6 5 4

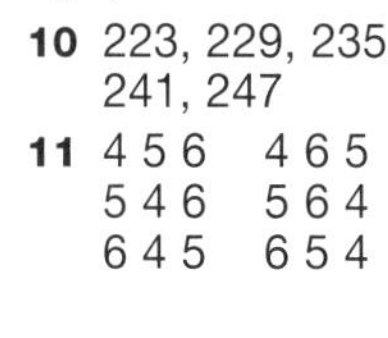

Space

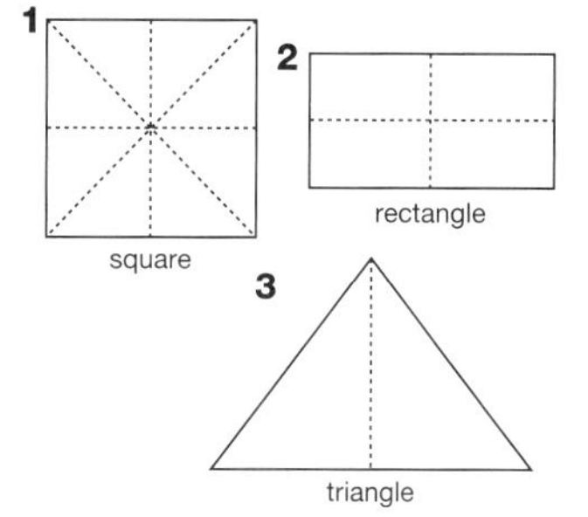

Probability

1 Likely is the best choice in this instance.
2 Hands on

UNIT 20 Number and Algebra

SET 1

1 27
2 29
3 32
4 35
5 54
6 14
7 10
8 14
9 15
10 12
11 27
12 12
13 90
14 75
15 $14

SET 2

1 24
2 8
3 40
4 8
5 10
6 40
7 21
8 7
9 ice-cream
10 3
11 12

SET 3

1 45
2 80
3 $33
4 $103
5 21 children
6 13
7 31
8 39
9 6

SET 4

1 60
2 5
3 40
4 $4.73
5 80
6 1, 2, 3, 4, 6, 12
7 2720
8 $9
9 $1.70
10 110
11 Hands on

Number and Algebra

Hands on. Some possible combinations.

	$2	$1	50c	20c	10c	5c
$8.95	IIII		I	II		I
	IIII			IIII	I	I
	III	II	I	I	II	I
	III	I	III	II		I

Space

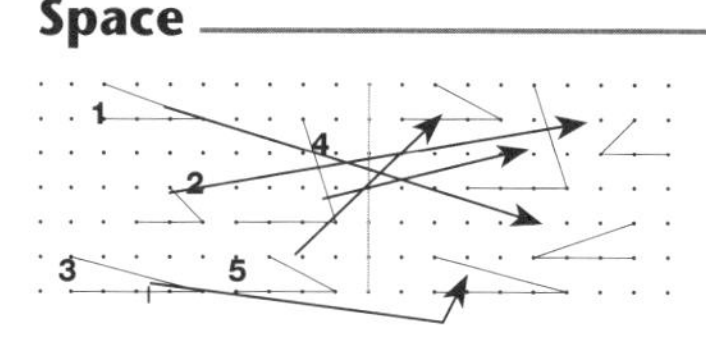
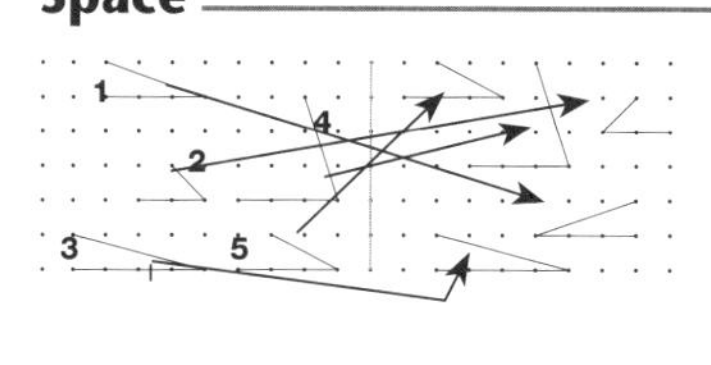

UNIT 21 Number and Algebra

SET 1

1 3
2 18
3 40
4 20
5 42
6 15
7 42
8 32
9 10c
10 2
11 59
12 40
13 13
14 31
15 11

SET 2

1 403
2 311
3 224
4 231
5 241
6 333
7 620
8 720
9 545
10 Liz
11 $91

SET 3

1 $60
2 $90
3 $60
4 $90
5 $70
6 $80
7 $40
8 $60

SET 4

1 2
2 0
3 6
4 87, 90
5 $4
6 $20 each
7 Yes
8 23
9 7000
10 2000
11 300
12 60c
13 12
14 232, 234, 236, 238

Space

Measurement

Answers

UNIT 22 Number and Algebra

SET 1

1 9
2 24
3 31
4 41
5 24
6 10
7 19
8 18
9 24
10 18
11 30
12 60
13 24
14 29
15 21

SET 2

1 594
2 683
3 782
4 885
5 893
6 883
7 872
8 884
9 673
10 475 children
11 $882

SET 3

1 5
2 8
3 9
4 3
5 6
6 9
7 7
8 8
9 5
10 7
11 3
12 4
13 3 passengers
14 6 eggs
15 5 sheep

SET 4

1 February
2 61
3 43, 46
4 93
5 10 o'clock
6 250
7 24
8 $5.30
9 $1.85c
10 8 each
11 72
12 48
13 11 pm
14 $\frac{1}{2}$
15 2000
16 26

Space

equilateral | isosceles | scalene

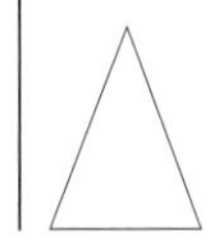

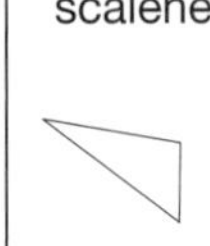

Measurement

1 1:30 2 4:15
3 7:50 4 2:25
5 4:40 6 10:10

UNIT 23 Number and Algebra

SET 1

1 12
2 100c
3 $6
4 5c
5 60c
6 5c
7 16
8 0
9 $17
10 10c
11 10c
12 27
13 $1.25
14 45
15 8

SET 2

1 427
2 527
3 329
4 326
5 319
6 328
7 228
8 216
9 Secret word COMPUTER

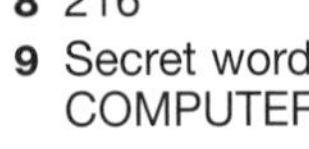

SET 3

$495 hands on including

	$100	$50	$20	$10	$5
1	IIII	I	II		I
2	III	II	IIII	I	I
3		卌IIII	II		I
4	IIII		III	III	I
5	III		卌IIII	I	I

6 $51

SET 4

1 $1.70
2 $1.85
3 $1.51
4 $4.33
5 50c each
6 300c
7 $2.70
8 90c
9 Apple pie
$5 – $1.30 = $3.70

Space

Hands on

Statistics

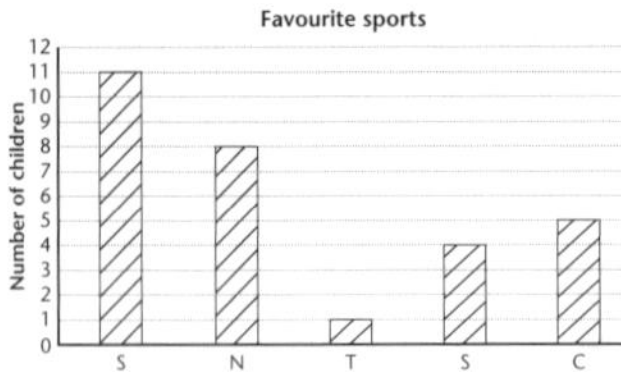

UNIT 24 Number and Algebra

SET 1

1 14
2 11
3 23
4 13
5 13
6 13
7 25
8 12
9 50
10 30
11 35
12 20
13 27
14 8
15 30

SET 2

1	23+38	23+40−2	61
2	32+29	32+30−1	61
3	18+49	18+50−1	67
4	25+39	25+40−1	64
5	27+48	27+50−2	75
6	46+19	46+20−1	65
7	48+49	48+50−1	97
8	53+28	53+30−2	81

9 58
10 79
11 95
12 83
13 108

SET 3

1 $\frac{1}{5}$
2 $\frac{3}{5}$
3 $\frac{2}{5}$
4 $\frac{4}{5}$
5 or any 3 parts
6 False
7 True
8 True

SET 4

1 $45
2 870
3 About 170
4 $3
5 270
6 Two hundred and seventy-two
7 $2.40
8 300
9 30
10 5
11 Hands on
Some solutions:
3 T-shirts
10 ties
6 socks
2 T-shirts, 2 pairs of socks
5 ties, 1 pair of socks, 1 T-shirt
1 T-shirt, 4 pairs of socks
5 ties, 3 pairs of socks

Statistics

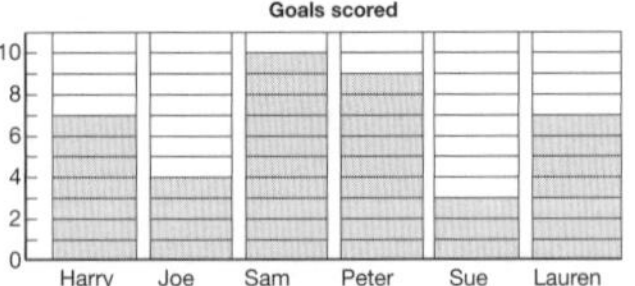

Measurement

1 months 2 minutes
3 minutes 4 seconds
5 hours 6 seconds
7 minutes 8 hours

Answers

UNIT 25 Number and Algebra

SET 1

1 25
2 0
3 39
4 15
5 1000 m
6 28
7 6
8 10
9 20
10 90
11 36
12 40
13 2
14 10
15 $18

SET 2

1 13, 15, 17, 19, 21, 23, 25, 27
2 5, 15, 25, 35, 45, 55, 65, 75
3 90, 85, 80, 75, 70, 65, 60, 55
4 $\frac{10}{10}$, $\frac{9}{10}$, $\frac{8}{10}$, $\frac{7}{10}$, $\frac{6}{10}$, $\frac{5}{10}$, $\frac{4}{10}$, $\frac{3}{10}$
5 48, 42, 36, 30, 24, 18, 12, 6
6 6, 12, 18, 24, 30, 36, 42, 48
7 Hands on
8 Hands on
9 Hands on

SET 3

1 3
2 5
3 6
4 2
5 4
6 2
7 3
8 4
9 5

SET 4

1 36
2 250
3 15
4 5
5 18
6 48
7 4 each
8 42
9 189
10 $20 each
11 65
12 48
13 366
14 $17.00
15 2 km
16 145

Probability

1 yellow
2 blue
3 red and green
4 yes

Measurement

6 2 4 1 3 5

UNIT 26 Number and Algebra

SET 1

1 30
2 0
3 27
4 7
5 1000
6 18
7 8
8 6
9 17
10 89
11 24
12 31
13 2
14 10
15 61
16 3 × 7

SET 2

1 15
2 17
3 ×
4 −
5 5
6 ×
7 ÷
8 Hands on.
9 Hands on.
10 Hands on.
11 Hands on.
12 Hands on.

SET 3

1 691
2 692
3 865
4 892
5 981
6 684
7 782
8 595
9 736
10 $675
11 591 marbles

SET 4

1 45
2 250 g
3 4
4 5
5 18
6 20c
7 12 each
8 180
9 $\frac{1}{4}$
10 91
11 633
12 18
13 365
14 $64
15 2400

Space

Hands on.

1	2	3

Measurement

1 7:20 2 6:25
3 9:10 4 9:50
5 5:00 6 11:05

UNIT 27 Number and Algebra

SET 1

1 47
2 3
3 19
4 10c
5 40
6 Yes
7 20
8 32
9 149
10 39
11 12
12 Yes
13 11
14 24
15 18

SET 2

1 7
2 12
3 9
4 20
5 25
6 18
7 28
8 30
9 60
10 24
11 32
12 40
13 90
14 0
15

20	45	10
15	25	35
40	5	30

75

16 Yes

SET 3

1 10
2 100
3 16
4 160
5 12
6 120
7 21
8 210
9 20
10 200
11 35
12 350
13 18
14 180
15 $240

SET 4

1 $5.80
2 202 km
3 20 hours
4 171 cm
5 $7.50
6 $119
7 Possible solution

1 (top), 2 – 5 – 6 (middle row), 7 (bottom)

Space

1 6
2 5
3 6
4 8
5 8
6 12
7 18
8 8
9 5
10 10

Probability

Hands on. Possible solutions:

- 5 red marbles, 4 green and 0 blue
- 6 red marbles, 3 green and 0 blue
- 7 red marbles, 2 green and 0 blue
- 8 red marbles 1 red and 0 blue

Answers

UNIT 28 Number and Algebra

SET 1

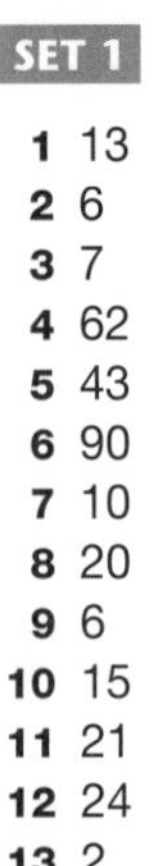

1 13
2 6
3 7
4 62
5 43
6 90
7 10
8 20
9 6
10 15
11 21
12 24
13 2
14 10
15 6

SET 2

1 4354
2 2684
3

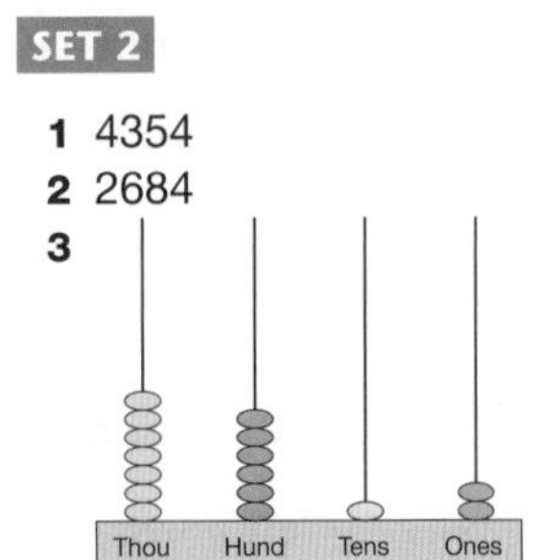

4 7618
5

Thou Hund Tens Ones

SET 3

1 15, 15
2 15, 15
3 25, 25
4 21, 21
5 35, 35
6 23, 23
7 38, 38
8 69, 69
9 No
10 10, 10
11 12, 12
12 15, 15
13 20, 20
14 35, 35
15 24, 24
16 No

SET 4

1 4
2 30c
3 73, 77
4 $12
5 No
6 24
7 177
8 83, 85, 87, 89
9 750 cm
10 471
11 4
12 $2.86
13 1, 2, 4, 8, 16
14 16
15 $5.40
16 6 r 3

Space

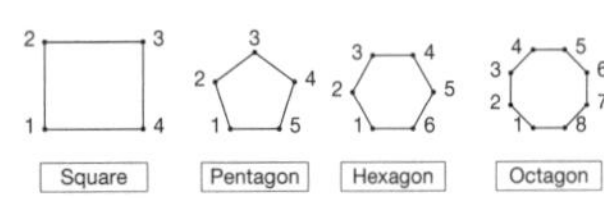

Measurement

2 kg 3
35 kg 5

UNIT 29 Number and Algebra

SET 1

1 7
2 5c
3 9
4 24
5 12
6 11
7 36
8 70
9 50
10 29
11 15c
12 28
13 500c
14 7
15 60

SET 2

1	35	40	45	50	55	60	65
2	80	75	70	65	60	55	50
3	93	88	83	78	73	68	63
4	10	17	24	31	38	45	52
5	0	15	30	45	60	75	90
6	145	135	125	115	105	95	85
7	150	135	120	105	90	75	60
8	0	30	60	90	120	150	180

SET 3

1 4 marbles
2 3 marbles
3 2 marbles
4 2 marbles
5 4 keys
6 2 keys
7 6 keys
8 5 keys

SET 4

1 786
2 991
3 106
4 317
5 $\frac{1}{8}$
6 $\frac{1}{2}$
7 400, 425
8 5 — 10, 40 — 20
9 1 — 4, 64 — 16
10 Hands on

Measurement

Ten past seven in the morning — 7:10 am
Half past eight in the evening — 8:30 pm
Twenty past eleven in the morning — 11:20 am
Twenty to five in the afternoon — 4:40 pm

Probability

1 Yellow
2 Yes
3 a Blue b yes

UNIT 30 Number and Algebra

SET 1

1 6
2 $13
3 20
4 30
5 18
6 33
7 16
8 36
9 21
10 45
11 32
12 24
13 18
14 36
15 5

SET 2

1 618
2 625
3 246
4 385
5 375
6 393
7 508
8 918
9 $522
10 $677

SET 3

1 266 km
2 313 km
3 283 km
4 292 km
5 405 km
6 281 runs
7 $88
8 51 eggs

SET 4

1 40
2 600
3 3
4 6
5 4 hund, 0 tens, 7 ones
6 321
7 91
8 50c and 20c
9 $1.05
10 7
11 3, 2
12 3, 4
13 5, 10
14 4, 9
15 5, 8

Measurement

1 200 mL
2 1000 mL
3 25 mL
4 120 mL
5 1500 mL
6 625 mL
7 2500 mL
8 3000 mL

Probability

1 one in six chance
2 one in two chance
3 certain
4 very likely

UNIT 31 Number and Algebra

SET 1

1 8
2 22
3 40
4 6
5 18
6 42
7 24
8 45
9 24
10 25
11 96
12 9
13 17
14 6
15 876

SET 2

1 1000
2 150
3 4006
4 7 tens
5 6 thousands
6 1234, 1341, 2341
7 653
8 149
9 4511
10 Hands on 6509

SET 3

SET 4

1 734
2 707
3 227
4 28
5 103
6 $9.07
7 7 r 1
8 32
9 $27
10 60
11 Yes
12 $7.30
13 $1.50
14 391
15 Feathers
16 Yes

Space

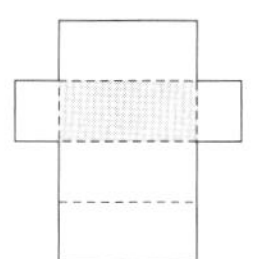

Measurement

1 9:54
2 10:26
3 5 mins
4 9 mins
5 10:08

UNIT 32 Number and Algebra

SET 1

1 38
2 21
3 13
4 18
5 5
6 15
7 32
8 19
9 41
10 33
11 42
12 10
13 25
14 21
15 19

SET 2

1 Hands on-some examples

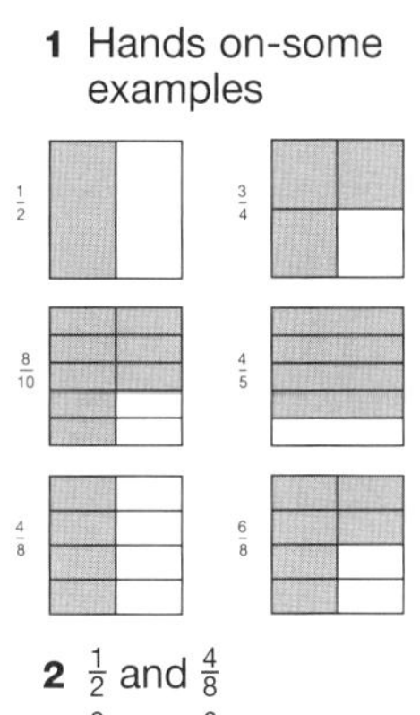

2 $\frac{1}{2}$ and $\frac{4}{8}$
$\frac{3}{4}$ and $\frac{6}{8}$
$\frac{8}{10}$ and $\frac{4}{5}$

SET 3

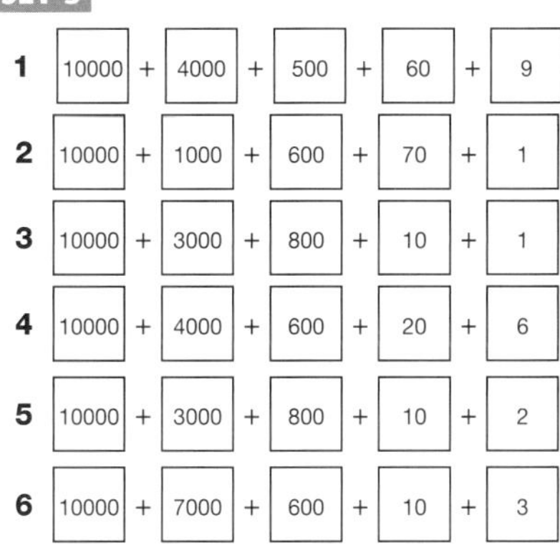

1 10000 + 4000 + 500 + 60 + 9
2 10000 + 1000 + 600 + 70 + 1
3 10000 + 3000 + 800 + 10 + 1
4 10000 + 4000 + 600 + 20 + 6
5 10000 + 3000 + 800 + 10 + 2
6 10000 + 7000 + 600 + 10 + 3
7 8510, 9671, 10811
8 11181, 11781, 11817
9 18621, 16634, 13671
10 16121, 15621, 15261

SET 4

1 20 days
2 8000
3 520
4 4
5 $3
6 3000
7 397
8 163
9 326
10 $3.78
11 $18
12 3580
13 300
14 45c
15 3
16 475

Statistics

1 No
2 No
3 No

Measurement

1 2
2 4
3 10
4 8

UNIT 33 Number and Algebra

SET 1

1 22
2 90c
3 36
4 5 tens, 7 ones
5 18
6 16
7 3
8 1999
9 One hundred and fifty-two
10 347
11 172 cents
12 50c
13 16
14 36
15 1357

SET 2

1 6
2 3
3 3
4 12
5 5
6 6
7 5
8 9
9 6
10 6
11 5
12 4
13 8
14 5
15 4
16 7
17 7
18 9

SET 3

1 133
2 232
3 227
4 337
5 606
6 662
7 578
8 949
9 2070
10 2384

SET 4

1 17
2 7000 g
3 4 × 5
4 $\frac{4}{10}$
5 1141
6 150
7 0
8 1250
9 458
10 32
11 4 r 1
12 28
13 367 + 582 = 949
Hands on

Space

1–7

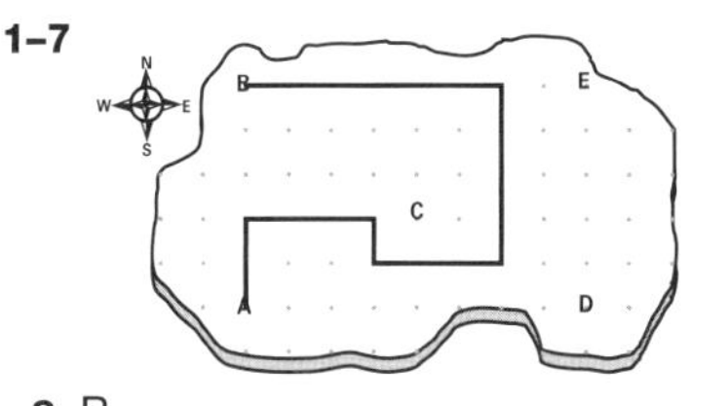

8 B

Measurement

1 Hands on

Answers

UNIT 34 Number and Algebra

SET 1

1 $80
2 2004
3 40
4 30
5 8
6 29
7 25
8 32
9 35
10 47
11 $30
12 $15
13 5569

SET 2

1 593
2 498
3 492
4 837
5 948
6 326

7

9	4	5
2	6	10
7	8	3

8

10	5	6
3	7	11
8	9	4

9

9	4	11
10	8	6
5	12	7

10

14	9	10
7	11	15
12	13	8

SET 3

1 triangle
2 circle
3 rectangle
4 arrow
5 hexagon
6 heart
7 star
8 diamond
9 pentagon
10 square

SET 4

1 6c each
2 30
3 22
4 300
5 $8.50
6 two hundred and sixty-seven
7 3
8 1336
9 20
10 $\frac{1}{3}$
11 306, 309
12 448
13 6 r 3
14 $24
15 $28
16 $20

Statistics

1 5
2 4
3 4
4 6
5

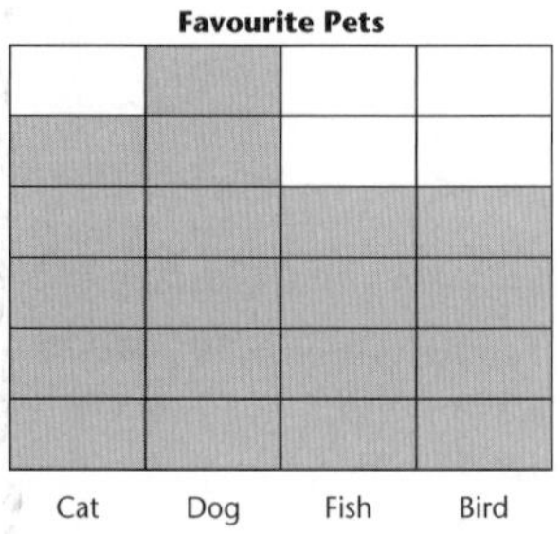

Favourite Pets

Cat Dog Fish Bird

6 19

Measurement

1 metres
2 centimetres
3 grams
4 centimetres
5 litres
6 millilitres
7 metres

UNIT 35 Number and Algebra

SET 1

1 17
2 19
3 12
4 9
5 40
6 30
7 21
8 16
9 61
10 15
11 30
12 120
13 22
14 12
15 18 km

SET 2

Hands on. Some examples

1 50 − 5
2 10 × 2
3 40 + 10
4 30 − 13
5 60 − 25
6 80 + 10
7 5 ÷ 1
8 90 − 20
9 15 × 2

SET 3

1 54
2 174
3 135
4 375
5 117
6 310
7 $60
8 $15

SET 4

1 842
2 35
3 $20
4 24
5 32
6 476
7 527
8 400
9 27
10 6
11 $40
12 13
13 Seven hundred and ninety-seven
14 220
15 240
16 180
17 $5

Space

1 Start at A. Travel east 4 spaces.
2 Travel north 3 spaces.
3 Travel west 2 spaces.
4 Travel north 2 spaces.
5 Travel east 7 spaces.
6 Travel south 5 spaces.

Measurement

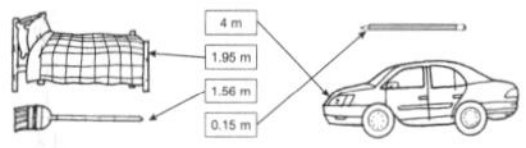

4 m
1.95 m
1.56 m
0.15 m